SCOTCH ALE

GREGORY J. NOONAN

A Brewers Publications Book

Scotch Ale
By Gregory J. Noonan
Classic Beer Style Series
Edited by Phil Rice

ISBN 0-937381-35-7
Printed in the United States of America
10 9 8 7 6 5 4 3

Published by Brewers Publications,
a division of the Association of Brewers.
PO Box 1679, Boulder, Colorado 80306-1679 USA
Tel. (303) 447-0816 • FAX (303) 447-2825

Direct all inquiries/orders to the above address.

Cover design by Robert L. Schram
Cover photography by Michael Lichter, Michael Lichter Photography
Cover art direction by Susie Marcus

Table of Contents

Acknowledgements

The author would like to give his thanks to the following people who made significant contributions to this book:

Mary Bertram, Caledonian Brewery, Edinburgh, Scotland
Dr. David Brown, Scottish & Newcastle Brewery, Edinburgh, Scotland
Ronald Borzuscki, Rose Street Brewery, Edinburgh, Scotland
Ian Cameron, Traquair House Brewery, Innerleithen, Scotland
Charles Finkel, Merchant du Vin, Seattle, Washington, U.S.A.
Dr. David Johnstone, Tennent Brewery, Glasgow, Scotland
Alan Hogg, Currie, Scotland
Hope and Nora Hogg, Currie, Scotland
Ronald Hogg, Edinburgh, Scotland
Jack Horne, Springfield, Massachusetts, U.S.A.
George Insill, Edinburgh, Scotland
Duncan Kellock, Maclay's Thistle Brewery, Alloa, Scotland
Roger Martin, Hugh Baird & Sons, Essex, England
Charles McMaster, Leigh, Scotland
Mary Nevins, Burlington, Vermont, U.S.A.
Adrian Newman, Belhaven Brewery, Dunbar, Scotland
Nancy Noonan, Burlington, Vermont, U.S.A.

Scotch Ale

Lady Catherine Maxwell Stewart, Traquair House Innerleithen, Scotland
Russell Sharp, Caledonian Brewery, Edinburgh, Scotland
Elizabeth Horne Wilcox, Charlemont, Massachusetts, U.S.A.
James Younger, Broughton Brewery, Broughton, Scotland

Dedication

To George Insill, maltster for McEwan's for more than 40 years, whose recollections of the Edinburgh breweries were invaluable to this effort, and whose preservation of records from these same breweries forms the core of the Scottish Brewing Archives, to Charles Mc Master, archivist and pub guide, and to the brewers of Scotland.

OH, GUDE ALE COMES

Oh, gude ale comes and gude ale goes,
Gude ale gars me sell my hose,
Sell my hose, and pawn my shoon,
Gude ale keeps my heart aboon.

I had sax owsen in a pleugh,
They drew a' weel eneugh;
I sell'd them a' just ane by ane,
Gude ale keeps my heart aboon.

Gude ale hauds me bare and busy,
Gars me moop wi' the servant hizzie

Stand i' the stool when I hae done,
Gude ale keeps my heart aboon.

Oh, gude ale comes and gude ale goes,
Gude ale gars me sell my hose,
Sell my hose, and pawn my shoon,
Gude ale keeps my heart aboon.

Rabbie Burns[2]

About the Author

Greg Noonan is the brewmaster at the Vermont Pub & Brewery and author of *Brewing Lager Beer* (Brewers Publications, 1986). He has written numerous articles in brewing periodicals, including a series on beer styles for *The New Brewer*.

Greg dates his interest in world beer styles back to savoring the limited imported beers available in the early 1970s. While hitchhiking across the United States in 1973, he sought out regional brews, but lamented their lack of character. The first homebrews he tasted were unearthed from a dirt root cellar where the corked and wax-sealed bottles had probably laid since prohibition. The sherrylike flavor of the sedimented beer was beyond anything that he had previously encountered.

The products of a homebrewing acquaintance piqued his interest even further, but his first attempt to brew his own did not follow until 1977. Greg dove in head-first, mashing his own grains on the very first brew. All-grain brewing was a subject that homebrewing literature of the time addressed only tentatively. Faced with an information void, Greg turned to professional literature for guidance.

That research and his concurrent experimentation and experiences led him to document what he learned in what has come to be a standard reference manual for home and craft brewers.

Greg and his wife, Nancy, opened their 14-barrel brewery and pub in Burlington, Vt., in 1988, after three years of lobbying the Vermont legislature to allow on-premises brewing. They chose to open a brewpub because it offered the best opportunity to brew not just a few brands, but a broad range of traditional beer styles. They eschew generic Amber and Gold appellations in favor of brewing to historic styles: Scotch and Irish ale, Best Bitter, Bavarian Weizen, Bohemian Pilsener and Vienna lager, to name a few of the 30-odd beer styles they have recreated in the past four years. The annual March tapping of their January 1st batch of Wee Heavy has become a *cause celibre* for the pub's patrons.

Although the demands of the brewery and the constant need for a busboy in the pub leave him little free time, Greg spends what leisure he can carve out with his family, reading (Kotzwinkle, Steinbeck, Burgess, Gardner) and drawing, as well as making the occasional pub crawl.

Scotch and Scottish Ales

ALLOA ALES

Awa' wi black brandy, red rum and blue whiskey
An' bring me the liquor brown as a nut;
O! Alloa Ale ye can make a chiel frisky,
Brisk, faewming a' fresh frae the bottle or butt.
An awa wi' your wines - they are dull as moss water,
Wi' blude colour'd blushes, or purple, or pale;
Guid folks gif ye wish to get fairer and fatter,
They aye weet your seasans wi' Alloa Ale!

Gif ye wish healthie habits an' wad be lang livers,
Then spiritous drinks ye s'ould never fash wi';
But Alloa Ale ye may drink it in rivers,
An' the deeper ye drink, aye the better ye'll be,
Sae potent as physic its virtues are valued,
They daily wha drink look hearty an' hale;
O ye a' hae heard tell o' a Balm got in Gilead,
Tak my word for't t'was neathing but Alloa Ale!

Then countrymen croud roun' the bizzing ale bicker,
An waur no on whisky your siller an' sense;
Nae gate ye'll fa' wi' the like o' this liquor,
That thro' body and saul can sic vigour dispense.
Let nae Brandy-bibber scare you wi' his scoffin,
At prudence in drink - till he tire lat him rail;
Ilk a dram that he drinks is a nail in his coffin,
But you'll lenthen your life-lease wi' Alloa Ale.

John Imlah, 1827[1]

Introduction

When I began to work on this book, I was doubtful about the value of any book being entirely devoted to a single beer style, and especially so for a style which is neither especially well known or documented in America. I thought the series was frivolous. I guess I was just dumber then. From the vantage point of having now seen four of the books of this series, and having been exposed to the wealth of information that I've been privilege to encounter in researching this book, today I have nothing but enthusiasm for the Classic Beer Style Series.

The previous books in the series cannot have been easy tasks for their authors. One great difficulty in looking at old beer styles is that no examples exist to do a sensory evaluation. Another is that records, if they survive at all, are generally so sketchy that little usable information can be extracted from them. As antiquarian beer researcher Dr. John Harrison notes, most pre-1850 brewery ledgers are "undecipherable."[3] Moreover, since the Middle Ages brewing has either been a very haphazard cottage industry, or was controlled by guilds that jealously guarded their craft. These have been succeeded by equally secretive commercial breweries, protecting their trade secrets.

Where old records do exist, they are more often curiosities than they are valid reference sources. Until the late 18th century brewers had no scientific means of monitoring and recording temperature. The fact that "eight quarters of malt were mashed with fifteen barrels of liquor" doesn't really tell us much of anything. Even after the thermometer and hydrometer came into use, brewery records still weren't very specific about such things as the volumes of beer that they were producing, or the color and nature of the malt used. Useful information about the type of hops, their condition and alpha acidity just do not exist at all.

Another set of problems arises with the weights and measures that are used in old records. The Scots used a measure called the boll for wheat and meal, and sometimes for malt. Although a boll (that's four firlots or eight auchlets) was nominally about the size of a U.K. bushel, it was variously two to six bushels in size as well. The standard for a U.K. bushel of malt is 42 pounds, but as late as 1847 an account describes "a bushel of the best Scottish malt weighing forty pounds."[4] A Scots pint was equal to three Imperial pints. While the size of a hogshead was standardized at 63 Wine Gallons (Wine Gallons are identical to U.S. gallons) in 1423, it was just as often 51 Imperial gallons. In London, a hogshead was 54 gallons for beer and 48 gallons for ale. Recorded weights and measures then, especially before 1850, are not of much use.

The Scots further complicated all this by valuing beer by Shillings-per-barrel and Guineas-per-hogshead, including excise tax (a guinea is 21 shillings—20 shillings make a pound sterling). As taxation changed (not only the tax rate, but whether it was malt, volume or density that was being taxed) and currency standards fluctuated, so the original gravity of ales sold at a given price changed. The strength of a 90 shilling ale was never fixed, and it is almost impossible to correlate all the periodic fluctuations to give meaning to the older records, especially since the containers themselves were not standardized. A 19th-century account laments the practice of quoting prices per hogshead, in that "a hogshead, which ought to contain fifty-four gallons, and run between twenty-seven to twenty-eight dozen, contains upwards of sixty gallons."[4]

More confusion is caused by nomenclature; while "Scotch Ale" means a Wee Heavy in Belgium, in the North of England it means a 70 shilling heavy, and a Wee Heavy is called a "Wee Dump." Modern brews employing non-traditional ingredients and technology make use of the old names arbitrarily.

It is with the Scotch ales of the 18th century that one would most like to begin. Even as early as 1837 the brewing methods and character of Scotch ale had changed, leading one writer to lament the passing of "the system which obtained in former times, when Scotch ale deservedly held, and still holds, the first rank amongst fermented liquors of British manufacture."[4] Fortunately that author recorded much about the old Scottish system; unfortunately, the character of the ales can only be estimated. No defining scientific records exist.

Even so, 19th-century brew records give none of the specific information that modern brewers are used to working with. There are records of hopping rates, but none defining the alpha acidity of the hops used. Bitterness levels, mineral analysis of the brewing water, degrees of malt color, pH and yeast character are not a matter of record.

Having made much of the inadequacy of old records, I must emphasize that it is the preserved ledgers of Scottish breweries and brewings, and journals of the brewers, that have made the compilation of this book possible. An appreciable number of these have been collected in the Scottish Brewing Archive, thanks to the diligence of maltster George Insill and other farsighted members of the Scottish brewing community. Mr. Insill's career spanned the years of the greatest upheaval that the breweries of Scotland, and indeed of most of the world, had ever experienced. Closings, buyouts, consolidations, contaminated wells and taxes caused the number of Scottish breweries to contract until only a handful remained. Mr. Insill saved the records from one brewery, then another and another. He lugged unwieldy ledgers from brewery to brewery, only to have many discarded by unthinking people in spite of his stewardship. The records that he saved eventually became the core of the Scottish Brewing Archive. Brewing ledgers, purchasing and sales records, photographs, notebooks and monographs have been preserved for future generations.

Scotch and Scottish ales provide good examples of beer styles that, outside of Scotia's shores, are widely misunderstood. Most of us have been only vaguely aware of what defines and delineates these beers. Like German alts, Irish ales, English milds and lagers of the Vienna style, there has been a vague awareness of the flavor and color parameters of the beers, but a lack of verifiable information that precisely defines those descriptors, and an almost total lack of information regarding the particular ingredients and nuances of processes that create those parameters.

It is my hope that this investigation of the Scotch and Scottish ale styles, as uninformed and lacking as it is, will adequately serve to better define them both, and allow brewers present and future to not only recreate these beers, but to learn from the processes and particulars of the styles to use ingredients in ways that will improve their brewing skills and repertoire.

This hope is the key to the enthusiasm that I have developed for the Classic Beer Style Series. Not only does it enable us, in our own time, to enjoy the research efforts of our fellows, and use the information to keep historical beer styles alive, but future generations will be privy to a broad spectrum of styles and historical practices in something more than bits and pieces of information. This series should provide firm footing for the research efforts of brewing fools not yet born.

As a final word, let me emphasize that this is NOT the definitive work on Scotch and Scottish ales. I trust that any information that my efforts left undiscovered will be compiled by other researchers to more fully define, categorize and document the historical evolution of Scotch and Scottish ales.

1

History of Brewing in Scotland

6500 B.C. TO 1820

Yes, this book is about Scotch. Scotch and Scottish ales, that is. Scotch Whiskey? It is nothing more than distilled Scottish ale. A historical beverage? Scotch whiskey did not even become popular until the early 19th century.[5] Beer, on the other hand, has been the Scots' choice among fermented beverages since Alban's earliest recorded history, and is still the beverage of choice in Scotland, accounting for 250 percent more of the alcohol consumed than all spirits combined.[6] This book will explain more fully the difference between Scotch ale and Scottish ales, but for a preliminary definition the term "Scotch ale" is reserved for the stronger brews (SG 1.070 to 1.130).

How old is brewing in Scotland? Archaeological evidence from circa 6500 B.C. indicates that the Picts were producing some sort of fermented beverage on the Isle of Rhum way back then.[7] Researchers believe that Picts and Scots were brewing barley beer before the Roman Empire invaded Britain in 43 A.D.[8] In the late fourth century, the Irish High King and adventurer, Niall of the Nine Hostages, undertook a genocidal campaign against the native Scottish Picts, at least partly to gain the secret of brewing heather ale. He was successful at killing off the Pictish population of Galloway, but he failed to extract the secrets of heather ale from even the last survivor. Heather ale survived, and was still brewed in the Orkneys until the early part of the 20th century.[5]

We are luckier than the marauding Niall. C. H. Cook, writing under the pseudonym J. Bickerdyke, recorded the manner of heather's use in his 1886 *Curiosities of Ale and Beer*: "The blossoms of the heather are carefully gathered and cleansed, then placed in the bottom of vessels. Wort of the ordinary kind is allowed to drain through the blossoms, and gains in its passage a peculiar and agreeable flavor known to all familiar with heather honey." He states that heather ale in the 15th century was probably two-thirds heather and one-third hops for bittering.[9]

As elsewhere in medieval Europe, monasteries were responsible for the first commercial breweries of scale. Alfred Barnard, in his three-volume *Noted Breweries of England and Ireland,* states that "the old monks of the Abbey of Holyrood were famed for their nut brown ale."[10] When St. Mungo established his monastic community in present-day Glasgow in 543 A.D., the monks brewed with the water of the Molendinar Burn, on the site where Glasgow Cathedral would be built in 1136. By the middle of the 13th century, Dominican Blackfriars and Franciscan Greyfriars were also brewing in Glasgow.

The Belhaven brewery, on the east coast at Dunbar, is built on the site of a monastery. It encloses a 13th century garden and two wells which date from the 15th century; these were used to supply water for brewing until 1972 (the

Label courtesy of
The Broughton Brewery.

house of retired brewer Sandy Hunter, adjacent to the garden, is known as Monkscroft).

"Browster wives" made domestic brewing a home industry during the Middle Ages. Bickerdyke records that "their ale was frequently made from barley and oat malt ... the lack of piquant flavor, afterward supplied by the hop, was in those days compensated for by the addition of ginger, pepper, spices and aromatic herbs."[9] With the beginning of the colonial trade in the 16th century, "howss" brewsters substituted molasses (treacle) for malt to brew treacle ale, also called brown robin and skeachen.[5]

"Publick" breweries existed in Scotland at least by 1488. In that year, it is chronicled that James the First of Scotland purchased Blackford ale in Perthshire, on the North slope of the Ochil hills. The three breweries there had abundant water good for brewing and malting barley available locally from the Vale of Strathallan. Blackford ales, in fact, were later to be famed throughout Scotland.[11] Secular breweries had sprung up in Glasgow as well. Drygate, Ladyvale, Gallowgate (Molendinar Burn) and Wellpark were sites of medieval breweries. The trade there was formally established by the Incorporation of Maltmen in the 16th century; most maltsters of the time were brewers as well.[5]

Well-aged and strong Scotch ales had gained an international reputation at least as early as 1578. They were familiar to the French since at least the 12th century, by virtue of the "Auld Alliance" that bound Scotland and France together against the expansionist aims of the English kings.

Trade was a two-way street, however, and in 1625 the Scots parliament forbade "the hamebringing of Foreyn Beir,"[5] that is, the importation of English ales and their Baltic counterparts (sowens).[11] Imports were a threat to the domestic brewers. Although their strong ales were the envy of England and the continent, their everyday ales were much maligned. Elsewhere in Europe the hop had become essential to the taste of "beer," but hops were still little used in Scotland. Hops could not be grown during the limited growing season therefore had to be imported at great expense. Common and table beers were insipid. When the ban on importation was universally ignored, Parliament reacted quickly, and in 1627 fixed the retail price-per-tun for imported ales at £6 to discourage their importation.[11]

Although Glasgow had its guild for malt*men*, the burgess (female) brewer was more common during the late medieval period and into the 17th century.[11] Although the guilds had, for the most part, replaced the browster wives in the southern towns, Aberdeen in the north still had 144 burgesses practicing their craft as late as 1693. Glasgow had 14 established breweries in 1700,[5] but brewing as a domestic industry of women survived at least until the end of the 18th century.[11]

The early part of the 18th century saw a renewal of the Auld Alliance, and with it a marked preference by the bourgeois for French wine over domestic ales. Wine was chic. Hopped English and continental ales also began impinging upon the brewers' market, as did the infant distilled-beverage industry.[11] At virtually the same time, the pastoral and agrarian base of the Scottish economy was being remolded by colonial trade and the start of industrialization in the South. The beginning of a wage-based economy, combined with urbanization and population growth brought about by a stabler economy fueled a cash-paying demand for alcoholic beverages.[8] These challenges are what in all likelihood spurred the Scottish brewers not only to improve their products, but to do so to such an extent that Scotch ale would reign at the apex of internationally acclaimed beers for more than a century.

Ian Donnachie, in *A History of the Brewing Industry in Scotland,* states, "the transformation of brewing from a domestic craft into a mass-production consumer industry was essentially a response to the rising living standards of a growing population." The growth was at first regional in nature; "most towns of any size had at least one brewery" serving accounts that could be "reached by the brewery cart-horse and waggon in an hour's haul." By the 1790s there were 150 public breweries in Scotland, where there had been less than 50 in 1750.[11]

The years of the Scottish industrial revolution (1780 to 1820) were also the years in which public breweries experienced their greatest growth. By 1820 there were 240 public breweries. Within 20 years, however, the same population growth and improved transportation that had fueled great brewery growth during the late 18th century would lead to the demise of rural breweries. Combined urbanization, industrialization and the opening of export markets, created market forces favoring urban breweries. By 1825 Edinburgh had 29 breweries, Glasgow 27 and Alloa six. Brewing soon became the domain of these three southern cites.[11]

Several factors determined this shift. The first was the proximity of those cities to the Carse of Forth, the East Neuk of Fife and East Lothian, "the granary of Scotland," which had been growing barley for malting and brewing since the Middle Ages.[13] Summer farm laborers provided the manpower for wintertime malting and brewing.

Proximity to ports was the second factor in the ascendency of these cities. Although the breweries of Falkirk, Stirling, Limekilns, Dumfermline, Cambus and Bathgate also profited by their nearness to the source of Scotland's best barley, they lacked deep water ports. The western port of Glasgow and the Eastern ports of Edinburgh and Alloa are ideally situated on tidal rivers, the Clyde and the Forth, respectively. Ports gave Alloa, Edinburgh and Glasgow the edge on their competitors in the export trade, which mushroomed after 1830. To supply the demand from expatriate merchants and tobacco planters in the West Indies and North America, ale helped fill the holds of outbound trading ships. Scotsmen made up the bulk of the garrisons that Great Britain sent to the far corners of the world as the empire was built.[11] Scotch ale was their drink of choice.

The third factor was that these cities were among the earliest to be industrialized, and had the benefit of having been the 17th-century centers of glassmaking, iron founding and coal mining. They also experienced the greatest population growth as Scotland's population doubled between 1780 and 1850.[11] Finally, these three cities were blessed with naturally cool cellar temperatures and abundant clean water of a constant temperature.

The industrial revolution made more of that water accessible. Breweries that lacked artesian wells had used pumps driven by horses at a turnstile. Steam powered pumps that produced greater volume at less cost, and drills that bored down to deeper and more abundant aquifers were made possible by Edinburgh inventor/entrepreneur James Watt.

Although the Canongate brewery still used a horse mill for grinding malt as late as 1819,[11] most of the urban breweries quickly harnessed the energy of combustion. Mechanization reduced heavy labor and improved quality. Mashing was no longer a Herculean task. Roberts records in 1847 that "in late years the mashing-machine has been adopted instead of oars, in almost every brewhouse in Edinburgh."[4] Foundries abounded in southern Scotland, and contributed to the breweries' modernization: "modern coolers, instead of being of wood as formerly, are of cast iron ... being better conductors of heat. ... Some brewers, who have the command of water ... cool by means of a spiral pipe which traverses a large body of cold water."[4]

The industrial revolution also brought with it a new age of science. The information explosion offered brewers the tools to eliminate much of the chance and circumstance that had dictated the course of their brewings: "About the year 1760 brewers began to use the thermometer as a guide to direct them in regulating their heats, both for mashing and for fermentation; but so skeptical were they as to its utility, or practical value, that it was several years before it was brought into general use."[4] Hydrometers were more readily accepted, perhaps because thermometers had earlier proved so useful. A

Ian Cameron measures the temperatures of the wort in the wort chiller at Traquair House Brewery. Photo by Greg Noonan.

hydrometer developed in 1768 by Mr. B. Marin of Fleet Street for distillers was further modified by Alexander Allen of Edinburgh for use by brewers.[4] Scots brewers quickly took to the use of the saccharometer after its pioneering in 1784 by J. Richardson. Measuring instruments also led to better record keeping in the breweries."[3] Record keeping fostered an analytical approach that continues to accompany brewing to this day. Still, however, brewers of the day lacked information that we consider commonplace. At this time brewers still believed that they were merely dissolving "Saccharum, Fecula and Gluten" (sugars, protein and starch) in the mash. There was no conception that these were actually being enzymatically reduced. They recognized that protein was required to help fermentation, but erroneously believed that starch caused overattenuation and spoilage.

The ale that was brewed in Edinburgh, Glasgow and Alloa was, by today's standards, almost unbelievably high in both original and terminal gravities. Whereas today's 90 shilling ale starts at an original gravity (OG) of 1.075 and is packaged at SG 1.016-20, a 90 shilling ale in the early 19th century had an OG as high as 1.125, and a final gravity (FG) of 1.055. It consisted of approximately nine percent alcohol by volume. It wasn't called "strong beer" for nothing! A "small beer" of the time would have been more familiar to our tastes, but for Victorian Britain and its colonies, small beer was common enough stuff, and "twopenny," the weakest ale, was drunk as we consume soft drinks. Until the 1800s, brewers did not commonly bottle for domestic markets, but sent their ale out from the brewery in barrels to the trade. Small beer was generally bottled and corked by public houses and sold for between nine pence and one shilling per dozen.[4] Presumably, strong ale would have cost up to twice as much as small beer cost, while twopenny was named for its late 18th-century price per Scots pint, a measure which is almost equalled a U.S. gallon.

Label courtesy of Caledonian Brewery.

Strong beer became relatively more popular even as overall production increased between 1780 and 1840. In 1787, a total of 24,000 U.K. barrels of strong beer was taxed, along with 108,000 barrels of small beer and 114,000 barrels of twopenny ale. In 1830, 111,000 barrels of strong beer were recorded by the excise office, while production of "table beer," the successor to small beer and twopenny, stayed nearly level at 229,000 barrels.

Whereas less than 5 percent was exported in 1780, by the early 19th century, about 10 percent of the strong beer was exported. The volumes and destinations of the exported beer were dictated by colonial expansion. In 1815 notable Scotch ale export markets were Jamaica and the West Indies, Canada, South America, Germany, the United States, Denmark and Russia; by 1850 the market had become the West Indies, East Indies, India, Australia, South America, Africa and Germany. "By 1815 Scottish exports were 13,700 barrels, over 60 percent being consigned to England, and a further 30 percent to customers in North America and the West Indies. ... In 1850 exports were 21,000 barrels. ... Asia, Australia and Africa ... together accounted for nearly half the total."[11]

1820 TO 1991

The popularity of the old Scotch Ales diminished at various times due to import restrictions, taxes (especially those that taxed beer by strength) and market shifts. The first market shift affecting Scottish brewers was the early 19th century rage for porter. Scots brewers, however, were canny; practical mercantilism has always been part of survival for the hardy Scots.

Breweries hired London brewers familiar with the brewing of porter and with its ingredients. The maltsters had to learn the techniques for producing brown porter malt, made by sprinkling the malt with water just before and during kilning to give it a slightly crystalline character. The malt was "blown" in the kiln; as the heat increased, the damp malt would swell and then pop, making the endosperm more readily soluble.[4] London porter was brewed with soft Thames river water, so circumstance favored the Edinburgh brewers, who were well blessed with abundant soft water. Porter enjoyed great popularity throughout the British Isles from 1780 until about 1820.

Scottish brewers' range of beers quickly expanded to include new beer styles as they emerged. While one Edinburgh brewer lists five Scottish ales for sale in 1819, Stein's Canongate brewery had a £2 and £5 ale, 30, 40, 80, 90, 115 shilling ales and porter as well.[11] The brewers learned new techniques with the introduction of new beers.

About 1820 there was a resurgence in the popularity of Scotch ales, starting, ironically enough, in London. Soon, Scotch ale was again traveling to Liverpool, Bristol and Dublin through the Forth and Clyde Canal (completed in 1792).[8] The 1820 prices in Bristol demonstrate the value attached to Scotch ale; it sold at a premium over other popular beers. Scotch ale sold at 11 shillings, Burton ale at 10 shillings, Bristol ale nine shillings and London porter for seven shillings. Ten years later in London, Edinburgh ale sold for the same amount as did Burton ale, at eight pence, while ordinary ale sold for six pence and London porter for four pence.[11]

Burton's Pale Ale would be the source of the next challenge to the export market of the Scottish brewers. A lighter beer that was more sparkling than the older styles and was heavily hopped with Kent Goldings had been brewed by Hodgson's brewery in London since the 1750s. It came to be called India ale, and later, pale ale.

Export to tropical colonies had proven its ability to stand up to the rigors of hot and humid climes. Previously, only very strong beers had successfully survived the long, rough voyage and adverse atmospheric conditions of the tropics. Their high alcohol content acted as a preservative, and their rich flavor masked the effects of oxidation. Hodgson discovered that hops would accomplish the same things. He brewed a lighter beer that was more satiating in the oppressive heat and could withstand the rigors of both voyage and climate. The Caribbean and Far Eastern colonies clamored for Hodgson's ale, at the expense of the export brewers of Scotland and of Burton-on-Trent in England.

The Indian and Asian colonies were England's newest and richest. "Up to about 1820, this trade was almost exclusively in the hands of ... Mr. Hodgson ... [His] beer was as well known in India, and as highly appreciated, as is London porter all over the world."[4]

Burton brewers began to imitate the Hodgson style about 1820, and their exports to India soon eclipsed Hodgson's. Burton ales arrived in the torrid zone in even better condition than Hodgson's, and had a flavor that was even more avidly sought. The extremely hard water of Burton was a natural match for a big hop character, and gave a fuller flavor to this lighter style. The success of India ale was also partly "in consequence of the Government now allowing the unlimited use of sugar in brewing;"[4] hard water allowed adjuncts to be liberally employed without producing a watery, insipid flavor. Cheap sugar from the equatorial colonies filled the holds of returning ships, payment for the hogsheads of ale that had lain there on the outbound voyage. It then became part of more ale for export and lowered its cost. By the 1840s, Allsop and Bass of Burton were exporting 10 times the amount of India ale as Hodgson. Hard water and hops replaced alcohol content and strength as the means of producing a keeping beer, and in consequence India pale ale became the world's first mass-market beer.[10]

As Hodgson's ale garnered fame abroad, returning merchants and soldiers who had developed a taste for "bitter" created a domestic demand for it as well. By 1837 Roberts would proclaim that "what is called India beer is now very generally used in Great Britain. ... It is of the greatest importance to those who wish to compete with others, that they should acquire every information regarding the method of making those beers." He goes on to caution, however, that "many who have brewed for this particular market have had good cause to repent doing so."[4] Brewers needed to learn the secrets of India ale.

The first was the use of paler malts. Emerging technology of the late 18th century made pale malts possible through improved kiln design, and the use of the nearly smokeless fuel, coke. The second secret was the use of adjuncts to reduce the final gravity. The third was the use of Kent hops, which had a finer aromatic character than varieties that previously had been used. The fourth was higher fermentation temperatures, which contributed to India ale's piquant bouquet. Fifth, hard water was required for its preserving and flavoring effects.

Several Scottish breweries enticed experienced India ale brewers from London and Burton to their employ. Especially in Edinburgh and Alloa, breweries had found that some wells they drilled produced ideal water for Scotch ales, while others gave too hard an edge to the malty strong beers. The former tapped soft water, and the latter, hard. These brewers already had the fifth necessity at their command; their hard water wells, previously cursed, became a blessing.

These lucky brewers, and they were many, started brewing the paler, weaker and more highly hopped India ales. Pale gradually eroded the overwhelming popularity of Scotch ale, first in the warmer climes of the empire, and then even in England and America. Moreover, increased hopping rates improved the keeping qualities of their lower gravity beers. Given their long experience with brewing keeping beers, the Scottish brewers were able to produce a more stable beer than most of their English competitors, even in the brewing of pale ales. Scottish brewers began "to set their tuns at a higher temperature, thereby hastening the process. ... This increased temperature, however, is still lower than that used by the English brewer."[4] The Scottish method of fermentation gave less harshness than the higher temperatures risked by the English.

India ale from Scottish breweries rivalled the fame of Burton ales, even to the far corners of the empire. Indeed, as retired maltster George Insill of Edinburgh points out, their hard-water, highly-hopped exports have confused the perception of traditional Scotch ales, because their pale ale became equally regarded and well-known. By the late 1880s, Scottish brewers were exporting more India pale ale and export stout than Scotch ale.[11]

Weaker, lighter, filtered and "sparkling" highly carbonated beers continued to displace the strong, dark, unfiltered, naturally conditioned traditional beers. Scottish brewers reacted by imitation, and so remained competitive in both domestic and export markets. They increased their historically low hop rates and used less malt, of a lighter color. By the 1880s sparkling lagers and ales were being brewed throughout Scotland, using brewing sugars to achieve the light color and body.

Bottles and bottling were very important to the competitiveness of the Scottish brewers. Beer kept better in bottles than in casks. An early bottle manufacturing center developed in the Forth valley until 1745, when a prohibitive excise tax was levied on glass. Still, however, bottling was more common in southern Scotland than elsewhere in Britain and on the Continent. Glassworks and bottling houses were more often than not separate enterprises, but Jacob Dixon combined both in the early 1800s at his Dumbarton Glass Work & Brewing Company. Although bottling continued to provide access to important markets for the Scottish brewers, it was not until mechanization replaced hand-blowing after 1875, coincidental with the demand for more highly carbonated, lighter and filtered beers, that bottling in Scotland began in earnest.[11]

The late 19th century was another period of technical and scientific advances. Although brewers still believed that nitrogenous matter "is principally the cause of Beer turning sour,"[10] yeast had come to be recognized as a living organism, and bacteria were starting to be identified. By the 1870s, many breweries were equipping laboratories and using them for quality control. In the 1880s Heriot-Watt University began to educate brewers, at about the same time that electrical power was introduced in breweries.[11]

Refrigeration was probably the most significant technological development in the history of brewing. It allowed brewers to chill their worts more rapidly, and with less exposure to the atmosphere. It made brewing and fermenting possible year round. Brewers could mature their beers in less than half the time it had previously taken. Refrigeration made artificial carbonation and filtering possible.

Brewers turned to carbon-dioxide injection to produce more sparkling beers. Pitch-lined casks, introduced with lager brewing, could contain more pressure than the traditional beeswax-lined tuns and puncheons, but could not compete with the faster turnover that artificial carbonation and refrigeration made possible. Moreover, the lighter pale ales and lagers did not require the long flavor maturation of the old Scotch ales, and did not suffer as appreciably from lack of conditioning. Scottish brewers embraced the new technology because it maintained their competitive position in the export trade. By 1900 they exported 123,000 barrels of beer annually.[11] Between 1870 and 1950, Scots brewers would account for roughly one third of all the beer exported from Britain.[13]

The employment of science and technology helped the

larger Scottish brewers survive in the world markets, and protect their domestic trade from foreign intrusions, but at the expense of the smaller brewers. Furthermore, "the widespread introduction of standardized mass produced beers in Scotland during the latter half of the 19th and the early part of the present century" had a decimating effect upon the diversity of beer styles and interpretations previously exhibited.[11]

The larger breweries went public in the late 1880s and the 1890s to finance modernization. Their increased output upset the domestic market balance, and doomed many of the smaller brewers. "Local variations in brewing technique and numerous distinctive local brews" soon became relics of the past.[11]

Charles Mc Master has catalogued the wide variety of beers that were being produced in the late 19th century: strong Scotch ales, table beers, dinner ales, three guinea ales, 90 shilling ales, special export ales, harvest ales, imperial stouts, double brown stouts, sparkling ales, India pale ales, nut brown ales, amber ales, invalid stouts, oatmeal stouts, Pilseners, golden lagers, Scotch porters and Munich lagers.[14] Today's range is anemic by comparison.

Other market forces would set the stage for diminished production of Scotch ales, and diminished diversity overall. A strong fundamentalist temperance movement that had been building in Scotland for more than 30 years became a political force to be reckoned with in the late 1880s. Brewers feared prohibition so strongly that more than one began to brew non-alcoholic beer, and others scrambled to diversify. Another tack was employed by many brewers; they legitimatized drinking beer by associating beer with temperance (as preferable to whiskey or gin drinking), and marketed their brews with "good for you" slogans. As Maclays' head brewer Duncan Kellock says, with a wry grin, "Doctors are people too. A doctor could be found that was willing to say anything for five pounds slipped into the back pocket. 'Very healthy, indeed, good for longevity,' and everybody was happy."[15]

Legislation in both England and Scotland earlier in the century had placed limits on the number of retail licenses. This not only made licenses themselves more valuable, but created a scenario in which tied houses guaranteed a brewer outlets for his production. Although tied houses had been common in England since mid-century, Scottish breweries did not follow suit until English brewers engaged in a frenzy of license purchases in the 1880s that threatened to shut out the Scottish brewers from sales south of the border. Scots brewers reacted by buying pubs in England, especially in their traditional stronghold in the northeast. Domestically, they initiated partial ties by making loans to publicans. After 1900, tied houses became more of a rule than an exception throughout Scotland.[11]

Traditional Scotch ale was also undergoing a transformation during this period. Not only was it no longer the brand leader beer of most breweries, but it was being "watered down" as well. Barnard records that "William Younger & Company's Edinburgh Ale was a potent fluid which almost glued the lips of the drinker together, and of which few therefore could dispatch more than a bottle. [They] still brew the celebrated Edinburgh ale, of a less potent quality, but their principal manufacture, India pale ale, is well known and appreciated in all parts of Great Britain as well as in foreign countries."[10]

The Nungate Brewery's 140 shilling ale at the turn of the century had an OG of only 1.084. Such a specific gravity would have been hard pressed to qualify as a 70 shilling ale a century earlier. This "devaluation" of the shilling system would continue. Restrictions limiting raw materials, output, and gravities during World War I diminished the character of the strong ales even further.[11] Dr. John Harrison

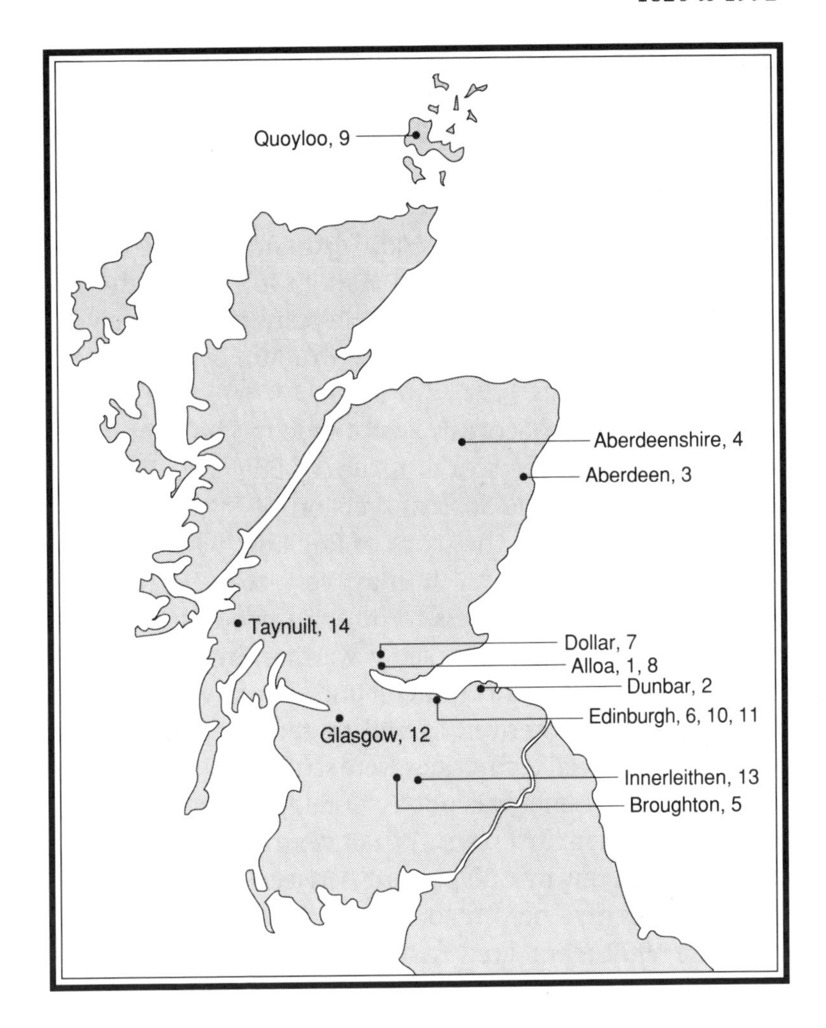

BREWERY MAP OF SCOTLAND

1) Alloa
2) Belhaven
3) The Bobbin Inn
4) Borve Brewhouse, Ruthven
5) Broughton
6) Caledonian
7) Harviestown Brewery
8) Maclay's
9) Orkney Brewery
10) Rose Street
11) Scottish & Newcastle
12) Tennent
13) Traquair House
14) West Highland Brewery

observes that "the essential difference between the 1871 and the early 1920s versions (of Youngers No. 1 Ale) was that the SG in 1871 was 1.102 [24 °B], whereas in 1923-24 it was 1.084 [20.4 °B].[3]

India pale ale displaced Scotch ale throughout Scotland, as it did "old ale" in England. Furthermore, at Tennent's Brewery in Glasgow and also at Heriot's in Edinburgh, and at Allsops in Alloa conversion of the plant to lager brewing satisfied a ready market such as would not be seen in England for decades.

On the whole, Scottish lager tended to be lighter and less flavorful than the sparkling ales of England. Lager, or more particularly, the Scottish version of Pilsener, would earn Scottish drinkers the scorn of England, which continues to the present day. It may very well be that the Englishman's pride has less to do with good taste than it does with geography. Markedly warmer English weather made lager brewing more difficult and expensive than it was for the Scots.[6] Moreover, as will be seen in chapter two, Scottish traditional techniques were so akin to lager practice that brewing Pilsener was not as "foreign" a concept to the Scots as it might at first seem. Whatever the case may be, the lager market grew by 500 percent between 1880 and 1885, held a steady five percent of the market share from 1900 until the 1960s,[9] and grew to account for 40 percent of the beer sold in Scotland in 1977.[11]

The 20th century saw the trend toward fewer breweries proceed by "rationalization" (i.e., takeovers and closures, in that peculiarly British propensity for euphemisms). Breweries disappeared first as beer sales slumped in the early 1900s,[11] then in the lean years between the world wars, and again in the takeover frenzy of the late fifties and sixties.

George Insill catalogued for me the changes that he saw during his life-long career malting barley for Edinburgh breweries.[15] Cask beers were the first casualty, losing popularity after World War I. To some extent returning service people had become used to bottled beers, and bottles were perceived as being more sophisticated. After World War II, artificially carbonated kegs, marketed as "brewery conditioned" draft, made their first impact on the pub trade. One of the motivating factors was the loss of skilled barmen and cellarmen during the war.

The quality of barley, both during the war and for some time after, was substandard. This encouraged the use of adjuncts, and contributed to lighter beers dominating the market.

The demand for fizzier beer by Americans stationed in Britain was also a factor; it is popularly believed that General Curtis Lemay introduced kegs to Britain, using the requisitioning might of the United States Air Force actually to provide the kegs to breweries so that they could be filled with highly carbonated beer for his airmen. By the 1960s, the keg and canned beer markets were growing, and bottling diminished accordingly. This led many brewers to abandon their bottling lines. The small breweries began contracting out their packaging. With the notable exception of Maclays and Belhaven, the beer went into cans.

After the war the British Empire began to crumble; former colonies levied increased duties on Scottish export beer. The taxes, combined with contraction in ocean-going shipping, led Scots brewers to turn again, albeit belatedly, to the acquisition of tied houses. In the 1950s free trade still comprised 65 percent of on-premises' sales in Scotland, but the trend to tied houses continued. In 1991, tied houses accounted for 75 percent of the on-premises' market. George Insill emphasizes that this fact has had a great effect upon the nature of beers being brewed, because the brewers and the large "independent" tied house chains can dictate to the customers what they are going to be served. Ale drinkers have lost the power that was inherent in free trade.

Although drinkers in the northeast of England and in Scotland retained a taste for the heavier traditional beers, the takeovers and closings of breweries after 1950 eliminated the diverse selection of Scottish brands. There had been more than 50 brands of Scottish ales from almost 20 brewers on the market in 1960,[11] but by 1991 there remained only six brewers, and less than twenty brands.

On the other hand, lager, which accounted for 54 percent of Scottish beer sales by the mid 1980s,[6] has experienced a falling market share since 1989. It has lost the sophisticated image that once made it popular.[13] Ironically, hoodlums are now termed "lager lugs" in the British press. One beneficiary of the changing tastes of younger Scottish drinkers has been cask-conditioned beer, which is showing an increase in sales. Although the larger breweries do not seem to be committed to its revival, all of the Scottish brewers offer at least one cask-conditioned Scottish ale, and the smaller breweries several. In the past decade, Scots have gotten a grip on their pints of real ale; cask-conditioned 60 shilling light, 70 shilling heavy and 80 shilling export flows again from the traditional tall fonts.

THE BREWERIES OF EDINBURGH

A serious look at Scottish brewing must certainly focus on Edinburgh. Edinburgh ale is almost synonymous with Wee Heavy (Scotch ale). Edinburgh is home to one of the three surviving small and independent 19th-century breweries; the Old Town looks today much as if nothing has changed since the last century.

Label courtesy of Caledonian Brewery.

Brewing is an ancient art in Edinburgh. "Publick" brewers were common there by the 15th century; in the late 16th century they formed a guild. The Fellowship and Society of Ale and Beer Brewers of the Burgh of Edinburgh (The Society of Brewers) was formed in 1575 to negotiate with the city fathers, to protect itself from foreign competition, to erect a common brew house, to set prices and to insure adequate supplies of barley and water.[11,14] The establishments of the Society of Brewers were all in the Old Town, bounded by Cowgate to the north, Candlemakers Row to the west, and Lothian Street to the south.

In 1598 the magistrates of Edinburgh ordered the Society to erect a windmill at the Boroughloch one-half mile to the south (drained in 1722, and called the Meadows ever since), in order to pump water from the loch to a reservoir in the Cowgate to supply their breweries.[14] This windmill alleviated the problem of the brewers running the town wells dry, but the solution was short lived. In 1618, alarmed by the falling level of the Boroughloch, the magistrates forbade the Old Town breweries the further privilege of using it as a water supply.[11]

Breweries had to bore their own wells to supply the copious amounts of water that they required. By good fortune, the wells of Cowgate, Holyrood and Canongate tapped a bountiful aquifer.[11] Dr. David Brown, Brewing Director for Scottish & Newcastle, sitting in his office at the old Morison's brewery on Horse Wynd, is a stone's throw from Holyrood Palace. From this vantage he points out that "the fault line from Holyrood palace up to the city gave clean pure water. Breweries literally ran up and down along the fault. Because the 51 degrees F [10.5 degrees C] temperature was an extreme constant, this also allowed for consistent mashing-in before the thermometers were used. After constant temperature, the purity and the softness of the water were its important attributes. Soft water lends itself to malty styles of beer, such as Edinburgh is famous for."[15] Breweries would follow this geographic fault out of the Old Town to Holyrood and Canongate to the east, thence to the North Back of Calton Hill (since renamed Calton Road) and later out along Fountainbridge and Slateford Road to the southwest, to continue to profit by its bountiful supply.

In 1710, a broadsheet entitled *The Brewers Farewell, to the Magistrates, Heritors, Merchants and Crafts of Edinburgh* complained that taxation and restrictions made the Old

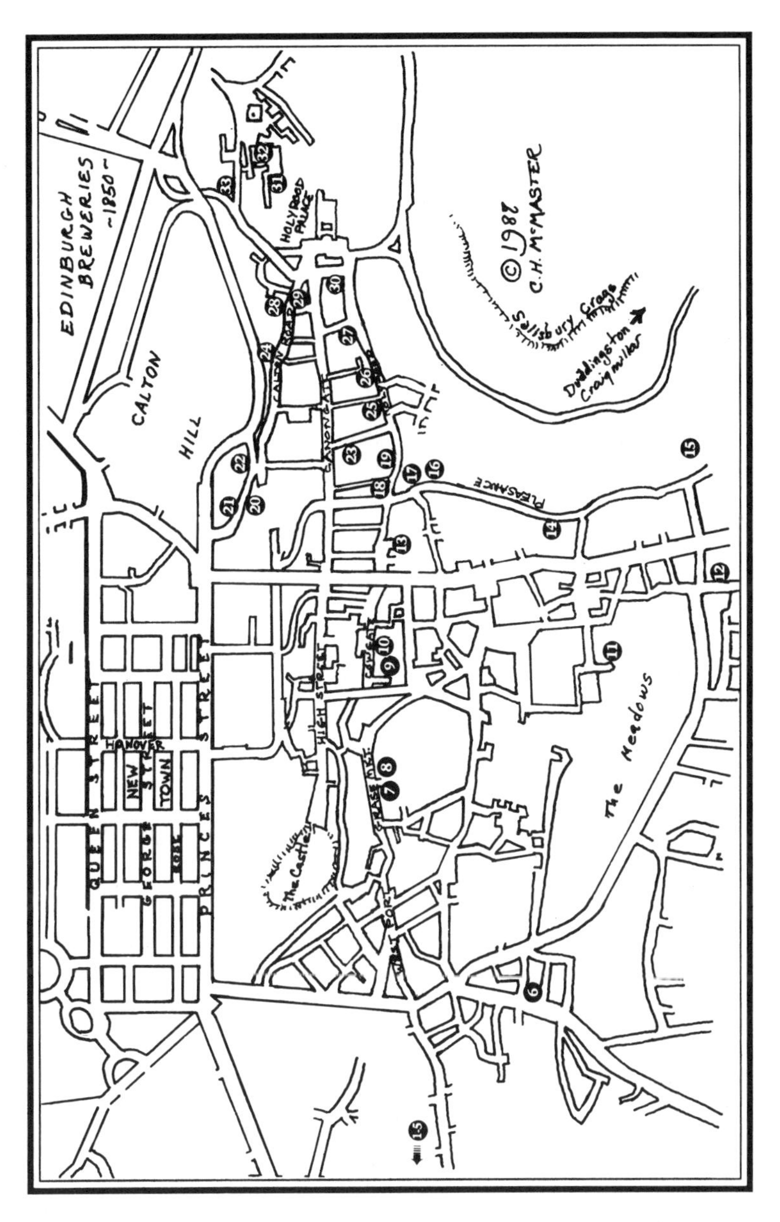

BREWERIES—EDINBURGH 1850 - 1970

EDINBURGH BREWERIES 1850 - 1970

KEY

1)	D. Bernard & Co.	1893-1895	Gorgie Brewery
	Bernards Ltd.	1895-1902	Wheatfield Road
2)	T & J Bernard	1889-1895	Edinburgh Brewery
	T & J Bernard Ltd.	1895-1960	(Slateford Road)
3)	Lorimer & Clark	1869-1920	Caledonian Brewery
	Lorimer & Clark Ltd.	1920-1970	Slateford Road
4)	John Jeffrey & Co.	1880-1934	Roseburn Brewery
	John Jeffrey & Co. Ltd.	1934-1960	(Heriot Brewery)
	Northern Breweries Ltd.	1960-1961	Roseburn Terrace
	Aitchison-Jeffrey Ltd.	1961-1962	
	United Caledonian Bwys. Ltd.	1962-1966	
	Tennent Caledonian Bwys. Ltd.	1966-1970	
5)	Wm. McEwan & Co.	1855-1889	Fountain Brewery
	Wm. McEwan & Co. Ltd.	1889-1959	Fountainbridge
	Scottish Brewers Ltd.	1959-1960	
	Scottish & Newcastle Bwys. Ltd.	1960-1970	
6)	Crease & Taylor	1850-1856	Drumdryan Brewery
	Taylor Anderson & Co.	1856-1870	Leven Street
	Taylor MacLeod & Co.	1870-c.1902	
7)	John Jeffrey & Co.	1850-1910	Heriot Brewery Grassmarket
8)	Cooper & MacLeod	1855-1910	Castle Brewery Grassmarket
9)	J & T Usher	1850-1860	Cowgate Brewery
	W & J Raeburn	1860-1897	Merchant Street
10)	Archd. Campbell & Co.	1850-1896	Argyle Brewery
	Campbell Hope & King Ltd.	1896-1970	Cowgate
11)	A. Melvin & Co.	1850-1907	Boroughloch Brewery
12)	Wm. Robertson & Co.	1850-1861	Summerhall Brewery
	Robin, McMillan & Co.	1861-c.1909	Summerhall
13)	Chas. Dick & Co.	1850-1869	
	Hope Bros. & Hart	1869-1872	Waverley Brewery
	Jamieson & Jenkinson	1872-1874	Cowgate
	Jas. Jamieson	1874-1887	
14)	Geo. Mackay & Co.	1867-1908	St. Leonards Brewery
	Geo. Mackay & Co. Ltd.	1908-1963	St. Leonards St.
15)	J & T Usher	1860-1895	Park Brewery
	Thos. Usher & Sons Ltd.	1895-1970	St. Leonards Street
16)	J. Fulton & Co.	1850-1909	Pleasance Brewery Pleasance
17)	Geo. Ritchie & Sons	1850-1889	Bells Brewery
	Edinburgh United Bwys. Ltd.	1889-1935	Pleasance
18)	Jas. Simson & Co.	1864-1896	St. Marys Brewery
	Simson & McPherson	1896-1901	South Back Canongate
19)	John Aitchison & Co.	1850-1895	Canongate Brewery
	John Aitchison & Co. Ltd.	1895-1959	
	Hammonds United Breweries Ltd.	1959-1960	South Back Canongate
	Northern Bwys. of Great Britain Ltd.	1959-1961	
20)	Andrew Drybrough & Co.	1850-1892	N. Back Canongate Brewery N. Back Canongate
21)	J & W Burnet	1850-1858	
	Jas. Steel & Co.	1858-1865	Craigend Brewery
	Steel Coulson & Co.	1865-1874	N. Back Canongate
	Drybrough & Co.	1874-1892	

22)	T & J Bernard	1850-1889	Old Edinburgh Brewery
	D. Bernard & Co.	1889-1893	N. Back Canongate
23)	Robt. Disher & Co.	1850-1889	Edinburgh & Leith Brewery
	Edinburgh United Breweries Ltd.	1889-1916	St. John Street
24)	Jas. Muir & Sons	1850-1883	Calton Hill Brewery
	Jas. Muir & Sons Ltd.	1883-1916	N. Back Canongate
25)	Morison & Thomson	1868-1878	Commercial Brewery
	J & J Morison	1878-1946	Canongate
	J & J Morison Ltd.	1946-1960	
26)	Alex. Berwick & Co.	1850-1858	Holyrood Brewery
	Wm. Younger & Co.	1858-1887	Canongate
	Wm. Younger & Co. Ltd.	1887-1959	
	Scottish Brewers Ltd.	1959-1960	
	Scottish & Newcastle Bwys. Ltd.	1960-1970	
27)	Wm. Berwick & Co.	1860-1869	St. Margarets Brewery
	Robt. Moyes & Co.	1869-1877	S. Back Canongate
	Moyes Brewery Co. Ltd.	1877-1878	
	Salisbury Crags Brewery Co. Ltd.	1878-1879	
	Douglas, Scott & Co.	1879-1880	
	Geo. Mackay & Co.	1880-1881	
28)	John Blair	1850-1873	Craigwell Brewery
	Chas. Blair & Co.	1873-1898	N. Back Canongate
	Gordon & Blair Ltd.	1898-1923	
	Gordon & Blair (1923) Ltd.	1923-1954	
29)	Thos. Carmichael & Co.	1855-c.1910	Balmoral Brewery
			N. Back Canongate
30)	Wm. Younger & Co.	1850-1887	Abbey Brewery
	Wm. Younger & Co. Ltd.	1887-1955	Canongate
31)	Robt. Younger & Co.	1850-1887	Abbey Brewery
	Wm. Younger & Co. Ltd.	1887-1955	Canongate
31)	Robt. Younger & Co.	1854-1896	St. Ann's Brewery
	Robt. Younger Ltd.	1896-1960	Croft-an-Righ
32)	City of Edinburgh Brewery Co. Ltd.	1866-1874	Croft-an-Righ Brewery
	Steel Coulson & Co.	1874-1888	Croft-an-Righ
	Steel Coulson & Co. Ltd.	1888-1960	
33)	J & G Pendreigh	1865-1870	Palace Brewery
	D. Nicolson & Son	1870-1889	Abbeymount
	Edinburgh United Breweries Ltd.	1889-1916	

Not plotted on this section of the map:

34)	Blyth & Camerson	1897	North British Brewery
	John Somerville & Co.	1897	
	Wm. Murray & Co. Ltd.	1922-1970	
35)	Drybrough & Co.	1892-1895	Craigmillar Brewery
	Drybrough & Co. Ltd.	1895-1970	
36)	Pattisons Ltd.	1895-1899	Duddingston New Brewery
	Robt. Deuchar Ltd.	1899-1954	Duddingston Road
	Newcastle Breweries Ltd.	1954-1966	
	Scottish & Newcastle Bwy. Ltd.	1960-1961	
37)	T.Y. Paterson Ltd.	1898-1936	Pentland Brewery
			Duddingston Road
38)	Wm. Murray & Co.	1885-	Craigmillar Brewery
	Wm. Murray & Co. Ltd.	-195?	Peffer Place
39)	J & G Maclachlan	1902-1907	Castle Brewery
	G & J Maclachlan Ltd.	1907-1947	Duddingston Road
	Maclachlans Ltd.	1947-1966	
40)	W & J Raeburn	1902-1913	Craigmillar New Brewery
	Robt. Younger Ltd.	1913-1931	Niddrie Mains Road
41)	John McNair & Co.	1850-1879	St. Anthony's Brewery
			St. Anthony Street, Leith

Town brewers unable to compete with the newer breweries that had sprung up around Holyrood Palace, outside the city walls. By the 1880s nearly 20 breweries and their maltings clustered about the palace.

Holyrood, built by James the Fifth in the mid 16th century, was the royal seat of the Stuarts. After the union of the crowns in 1603 it was the summer palace of the English kings. The St. Ann's brewery of Robert Younger Ltd. was "most immediately adjacent to Holyrood Palace, and which in fact shared a common wall with the latter on its north side." The area was "originally outwith ... Edinburgh's taxes and multures. So too with Canongate and Calton ... separate burghs. ... In addition, an abundance of good wells."[16] The Abbey Brewery, which Barnard calls "one of the most ancient in Britain," with portions dating to 1600, was owned by John Blair, a brewer to the palace.[10] Other breweries within the precincts of the palace were Croft-an-Righ (Gaelic for 'farm of the King'), the Palace, Holyrood, St. Margaret's, Morison's, St. Mary's, the Canongate.

The palace and the breweries were not always happy neighbors. "Queen Victoria, a periodic inhabitant of Edinburgh as a result of her sojourns at Holyrood Palace, was distinctly not amused by the smell and smoke emanating from the surrounding breweries, and was not slow to make her views known."[17] Edinburgh's civic planning report, The Abercrombie Plan of 1949, made plain that the city fathers did not consider brewing part of the future of Edinburgh. Regarding the cluster of breweries surrounding Holyrood, it stated that "there never was a case where greater conflict between property use existed. ... Royal Palaces and breweries cannot be said by any stretch of the imagination to make good neighbors. ... A Royal Palace shrouded in a pall of smoke, and the air permeated with factory processing, is no fit place for a sovereign to dwell."[17]

Water was the key factor in the clusters of breweries that developed in Edinburgh. The water determined the type of beer that Edinburgh brewers produced; strong, malt-predominant beer, given a deep amber-to-brown color by the malt used, by carmelization in the kettle and by the sheer amount of malt employed. At the Playhouse Close Brewery, Robert Disher brewed ten guinea ale in the early 19th century, "a very potent brew" of about OG 1.100 (26 °B), famous as the "Burgundy of Scotland".[10] Bell's Brewery at the foot of Pleasance, on the east side of St. Leonard's Wynd, brewed Black Cork, a strong Scotch ale that was one of Edinburgh's most popular until Robert Kier died in 1837, and took to his grave the secret method of brewing it.[18]

Edinburgh water was ideal for porter brewing as well, and so the brewers there not only weathered the heyday of porter, but made Scotch porter nearly as famous as that from London. Furthermore, wells that gave hard water and were of no use in the 18th and early 19th centuries would be put to good use from the 1820s on, for the brewing of India ale. The Croft-an-Righ Brewery had two interconnected wells, 200 feet deep, that gave 5,000 gallons per hour of hard water of excellent quality, and produced excellent pale ales. The same was true for breweries from Cowgate to Holyrood, and of the breweries built after 1850 to the west and southeast of the Old Town.

By the early 1800s a line of breweries followed the fault south, from John Miller's Potterrow Brewery below Canongate and Bell's at the foot of Pleasance, to James Anderson's (later Melvin's) 1575 Boroughloch, thence south-east to the Summerhall Brewery and James Kerr's Newington Brewery at Sciennes.[14] A quarter mile to the north of Holyrood Palace, Blair's Craighwell Brewery in Low Calton, and Dryborough's, Bernards, Muir's and Carmichael's breweries

Sign in Caledonian's collection, displayed in their employee's pub. Photo by Greg Noonan.

lined the North Back of Canongate at the base of Calton Hill.[18]

William McEwan was the first of the Edinburgh brewers to follow the fault past Grassmarket, building his Fountain brewery and maltings a mile west of Cowgate at Fountainbridge in 1856. It was a canny move, allowing him to take advantage of the transportation opportunities opened up by the new East-West rail line. A rail siding allowed the brewery's products to be loaded directly onto cars, cutting out the costs of intermediate shipment by draywagon. Eleven more breweries would follow the fault outward in the latter half of the 19th century; Lorimer & Clark's Caledonian, John Jeffrey's Roseburn (Heriot), T. & J. Bernard's Edinburgh, D. Bernard's Gorgie Breweries, Mackay's St. Leonard's, Usher's Park, Murray's Craigmillar, Deuchar's Duddingston, Paterson's Pentland, Maclachlan's Castle and

Raeburn's Craigmillar. New breweries stretched southeast past the Salisbury Crags to Duddingston Loch.

There were 30 breweries already operating in Edinburgh when Robert Clark, formerly the brewer at Boroughloch, and George Lorimer established their Caledonian Brewery on Slateford Road in 1869, extending the "charmed circle" of breweries tapping the fault. The quality of the well water at that point along the rail line determined the brewery's siting. The fault gave good water, variously hard or soft depending on the strata that the bore tapped, so that brewers enjoyed the versatility of being able to brew the soft Scotch and Scottish ales, or the sharper-edged India and pale ales.

With three breweries still operating, and numerous 18th and 19th century brewery buildings still standing, Edinburgh remains the brewing capital of Scotland.

ALLOA, "THE BURTON OF SCOTLAND"

Alloa's fame was built on the old dark strong Scotch ale, and sweet and mild ales. By the 1890s Alloa breweries were also brewing pale ale with hard water from the nearby Ochils; like the Edinburgh brewers, those of Alloa prospered in the late 19th century because access to hard water allowed them to remain competitive with Allsop and Bass of Burton.

George Younger's was the first Alloa Publick Brewery, in 1762. In 1764 he erected the Meadow Brewery. By 1890 their Candleriggs Brewery, purchased from Robert Meiklejohn in 1871, was the biggest brewery outside of Edinburgh, with two 150-barrel coppers, one 90—and one 45—.[8] Although George Younger of Alloa and William Younger of Edinburgh were not related, in the 19th century the two families' brewing legacies would eventually be united when William Younger's Brewery and William

McEwan's Brewery were united as Scottish & Newcastle in 1960. This merging has its roots in the marriage of George Younger's son James marrying Janet McEwan; their son George Viscount Younger, took over the Meadow Brewery at the age of 17 upon his father's death in 1868. His brother William Younger went into the employ of their uncle William McEwan in 1874, and succeeded McEwan upon his death in 1886.[19]

Alloa was ideally sited as a transportation hub. Eight breweries existed in 1845, brewing 80,000 barrels of ale.[8] What was not consumed locally was carted north and west, to Stirling and beyond, or was exported from the Alloa docks. George Younger was already exporting in the 1850s, principally "Demerara", a very strong ale that was matured in barrels before being bottled in stone bottles and shipped to the West Indies.[11] The bottled trade would be a major component of the Candlerigg's business in the late 1800s;[19] in fact, the export market was always a major outlet for the Alloa breweries. Other Alloa breweries of the 19th century in addition to Younger's were Blair & Company's Townhead, Calder's Shore, Maclay's Thistle, Henderson's Mills, Syme's Hutton Park, Knox's Forth Brewery at Cambus, Meikeljohn's Bass Crest Breweries (originally known as the Grange), and Arrol's Alloa Brewery.

Syme's brewery closed in the 19th century. The Shore Brewery stopped brewing in 1921, when all but one of its wells were contaminated by the collapse of the Forthbank Pit. Allsop's of Burton took over Arrol's Alloa brewery in 1920 and began lager brewing there in 1921. Bass of Burton took over Meikeljohn's in 1918 and stopped brewing there in the following year. The Mills Brewery discontinued brewing in 1941. The Forth Brewery was bought out in 1954 by Blair's for its tied houses, and the brewery was closed shortly thereafter. Blair & Company's

Townhead Brewery was closed up in 1959 as part of a buyout by George Younger's. Only the Alloa Brewery and the classic Thistle Brewery still survive.[8]

2

Scotch Ale

Scotch Ale is a strong 6 to 10 percent alcohol by volume (4.8 to 8 percent alcohol by weight), sweet and very full-bodied ale of 1.070 to 1.130 OG (17 to 34 °B), with malt and roast malt flavors predominating, of deep burnished-copper to brown color. Scottish Ale refers to ordinary Scottish ales, 60 shilling ale, 70 shilling ale and 80 shilling ale, ranging from 1.030 to 1.050 OG (5.7 to 12.8 °B), softly malty to slightly roasty, of burnished-copper to brown color.

Scots brewers did not share much of English practice. An examination of Scotch Ale is necessarily as much a dissection of process as it is an analysis of the product. Thanks to W. H. Roberts' timely preservation of early 19th-century Scottish methods and his numerous references to previous practices,[4] we know a great deal about how Scottish brewers brewed. Quoting liberally from *The Scottish Ale Brewer and Practical Maltster*, we can reconstruct typical Scottish practice.

Scottish practice was not only dissimilar from English methods in many respects, but strikingly similar to continental lager technique. David Johnstone, head brewer for Tennent's of Glasgow, credits Roberts for his observations that "the methods of continental lager production and Scotch ale production were virtually identical."[6] He summarizes Roberts' exposition of the techniques used by the Scots. (These methods were unique in the British Isles, deriving from the need to produce a "keeping" beer for the booming export trade). Johnstone's summary:

1. The use of sparging as opposed to the English technique of double mashing.
2. Boiling for one to 1 1/2 hours instead of three, to avoid spoiling the delicate aroma.
3. Low temperature fermentations (50 degrees F [10 degrees C]) lasting up to three weeks as compared to the English high temperatures (70 degrees F [21 degrees C]) fermentation of less than a week.
4. Storage in cold cellars for up to six months (lagering!)
5. The practice of skimming was rarely carried out—which would almost certainly result in a yeast strain with a tendency to bottom fermentation.
6. Brewing and fermenting in the winter months to facilitate cooling.
7. The use of pure, soft water.[6]

Roberts recorded that "it is not uncommon in Scotland for brewers to have their gyles in the tun for twenty-one days, whilst in England, so long a period even as six days is considered of rare occurrence."[4] Regarding India beers, he wrote that "the difference between the management of these beers at this stage, and that of ales made on the Scottish system [is that the Scottish ales are] conducted by a slow fermentation, whereas the India beers are carried through by a remarkably vigorous one, so much so, that the time the worts are in the fermenting tun, seldom exceeds 24 to 30 hours, and at an increase of temperature of about seven degrees. ... The pitching heat varies in different establishments from 58 degrees to 60 degrees ... with two pounds of yeast per barrel ... from November to March."[4] Scots pitched their worts in the vicinity of 50 degrees F (10 degrees C), with fermentation temperatures seldom rising above 65 degrees F (18 degrees C) or so.

Other factors also contributed to the distinctive character of Scotch ales, as Scottish & Newcastle's technical director Dr. David Brown points out. "Keep in mind that this is oversimplified, but the use of roast barley, and residual sweetness, determines more what makes a Scotch ale than does the yeast. The maltiness is from use of malt, not adjuncts. Traditional Scots malts were darker than English malts as well. We use a lot of malt, and up to two or three percent highly roasted barley, of 1200 °EBC [500 °SRM]. The maltiness is not from crystal. Roast barley accounts for flavor, rather than caramel malt as the English do. This is by a tradition that has stood for hundreds of years, rich and sweet beers.

"Another difference is that we don't prime our beers. Scottish ales are sweet not from primings, but from residual sweetness left in the beer by stopping fermentation, by using very sedimentary yeast that respond to temperature change. The yeasts are highly flocculant and only low attenuating. Low temperature fermentations produced low ester levels, another common attribute of Scottish ales. In modern practice, oxygen levels in the wort are controlled to provide the yeast just what they need for growth, to reduce the production of esters. Scottish yeasts produce very little amyl acetate [the banana ester], it's usually just above threshold. The lower growth also maximizes alcohol production. These are some of the things that are key to brewing Scottish ales."[15]

Modern Scottish brewers temper fermentation temperatures and drop the yeast out early to replicate past practices, which were undoubtedly predicated by the low prevailing temperatures. It is an unusually warm day in

Scotland when the average temperature rises above 50 degrees F (10 degrees C) between October and May. Modern brewers continue to ferment only to about one third gravity. Russell Sharp of the Caledonia Brewery points out that although the hop flavor is not generally evident in Scottish ales, hop bitterness plays an important supporting role, by masking this residual sweetness in Scotch and Scottish ales.

A taste of a Wee Heavy bears all this out. It is much like the flavor of a fine Scotch whiskey, moderately watered: malty. Scots whiskey is heady and fragrant, unlike corn and grain whiskeys. Single malts are sometimes slightly buttery and their alcohols are very aromatic, generally of an ethanol nature, with propanol forming part of their character. So with Scotch ale. The alcohol counterbalances and softens the sweet maltiness. The faint contribution of aldehydes formed from alcohols during aging contributes to a rich blend of flavors, so subtle that it is difficult to dissect them. Scotch and Scottish ales require extensive cellaring at cold temperatures, as was traditionally universal practice among Scottish brewers, for their flavor to develop. Scotch ales are invariably rich (even to the point of being syrupy) and mouth-filling. They are soft as well. Scotch ales show less acidity than is usual for beer; a sampling of Scotch ales gave a pH range of 4 to 4.75, whereas Scottish ales measured between 3.75 and 3.9, which is within the more usual range for beer.

There is a standard of color for the ales of Scotland. Dr. David Brown cites a color of 25 to 28 °EBC (approximately 12 °SRM) as usual for modern Scottish ales. George Insill considers 25 °EBC to be traditional as well, but notes that many ales were even darker. "Before the 1950s, Scots miners wouldn't drink a beer that wasn't of a full brown color. Customers demanded a color of 30 to 40 °EBC (approximately 15 to 20 °SRM), and they got it. Roast barley was used for color control; it also gave the characteristic bite to the flavor. It was good for heading, and didn't cause the infection and other problems that late addition of color additives did. Darker beers have always been the norm for primitive breweries, because lack of control alone, or what would be considered sloppy brewing by modern standards, will give a color of 18 to 20 °EBC. In Scotland, however, darker color is the tradition."[15]

With a look at ingredients and a further examination of process, we have enough information about Scottish and Scotch ales to not only brew in the style of this century, but to brew ales very similar to those that were famous world-over in the last century.

3

Water

"In the vicinity of Holyroodhouse in Edinburgh, some eminent brewers are to be found, who assert, that the excellence of their ale depends upon the water found within their premises; and further, that the water at a higher elevation is not capable of producing ale of equal quality."[4]

From high atop the basalt cliffs of the Edinburgh castle mound, one looks down into a city seemingly preserved in 19th-century garb. Princes Street and the New Town to the north has its 20th century facades, but even more of the architecture is of the 1800s. In the ravine to the south, Westport, Grassmarket and Cowgate wind toward Holyrood Palace amidst a medieval-looking jumble of buildings abutting narrow lanes (Candlemakers Row, Old Fishmarket Close, Horse Wynd). Below Calton Hill, rows of old brewery and malting buildings march toward Holyrood. Just beyond, the Salisbury Crags jut upward to Arthur's Seat, with Duddingston Loch just beyond. From any of the hills, it is obvious that Edinburgh is built along a geological fault, the lines of which run outward from Holyrood Palace through the city. One fault runs southwest through Cowgate, past Westport, and on to Fountainbridge and Gorgie, another west and north around Calton Hill, and a third south beneath Salisbury Crags to St. Leonard's, before broadening out at Duddingston and Craigmillar. The faults give access to a huge subterranean aquifer.

John Jeffrey's well book records the strata of an 1875 hole bored to deepen an existing well: From hard and soft white sandstone at 391 feet (119 meters), through thin strata of hard red sandstone, layered with red fireclay at 415 feet (127 meters), to the bottom of the bore at 500 feet (152 meters). Whereas clay and dense rock cap many an aquifer, white sandstone is nature's water filter; it produces clean and generally soft water. Another well was bored "through fireclay and fakes with limestone ribs as part of a cementstone group to 245 feet (75 meters), then Upper Old Red Sandstone from 985 feet to 1104 feet (300 to 337 meters). Lime and cementstone give water of varying degrees of carbonate character."[20]

Dr. Stephen Cribb observes in his article "Beer and Rocks" that "Edinburgh lies in the center of a heavily faulted, generally north-dipping pile of Lower Carboniferous and Upper Old Red Sandstone strata. The juxtaposition of different rock types has meant that individual breweries have always had access to differing sources of water. Even boreholes in close proximity could produce waters of vastly different analyses. Thus the Edinburgh brewers have always blended waters to produce their characteristically wide range of beers, from milds to bitters and beyond."[21]

The soft and malty style of Scotch and Scottish ales was in large part predicated by the softness of the water. The right water is a prerequisite of the style; the water should not give a bitter edge. Not even all the Edinburgh brewers had the right water.[8] Up until the mid 19th century, a well that gave hard water would have been considered poor by a Scottish brewer. It would not make a passable Scots ale.

When the market shifted to India ales later in the 19th century, hard water was absolutely required for its flavor and contribution to the beer's stability. The same hard-water bores that may have been written off as a loss in 1815 became prized assets by 1850, for the brewing of India ale. Edinburgh's brewers were thus blessed with the versatility that the city's geological underpinnings provided. Many breweries found hard and soft water side by side, at differing depths. Alloa had access to hard water as well. These two brewing centers were able to capitalize upon changing markets by using one or another well, or a blend of water from both. "The success of the Alloa and Edinburgh breweries was apparently due to the availability of varying water types which allowed brewing of different beers, depending upon which strata water was drawn from—as for Alloa, which brewed lagers from a well 700 feet (213 meters) deep, and ales from a shallow well."[8]

Before the mineral character of water was recognized as being an important commercial consideration, brewers valued water sources for two reasons: for freedom from organic contamination and for constant temperature. The first is not hard for us to understand, but the second is a bit more obscure, until one reflects upon the challenges faced by brewers before the invention of the thermometer. A constant temperature water supply meant that the number of variables affecting mash temperature was reduced. By keeping volumes and times the same, the constant-temperature water gave brewers repeatability. Edinburgh's well water gave forth at a very constant 51 degrees F [10.6 degrees C]. Over time, one hundred breweries would tap this source for its purity and consistent temperature.

Water from the aquifer beneath Edinburgh is consistent enough that its mineral nature is relatively the same from year to year, and is much as it was in past centuries. A representative sampling of water records from notebooks, analyses and well records preserved by the Scottish Brewing Archives gives a very detailed picture of the nature of the waters. Aitchinson's Brewery well (at St. John and Holyrood) was 233 feet 10 inches (71 meters) deep, gave 1,980 gallons per hour (75 hectoliters per hour) of water, at a hardness of 399 milligrams per liter as calcium carbonate ($CaCO_3$). John Jeffrey's Heriot Brewery in Grassmarket had wells 118 feet (36 meters) and 986 feet (300 meters) deep, that gave 3,600 gallons per hour (136 hectoliters per hour) of water with a total hardness of 266 milligrams per liter, and permanent hardness of 112 milligrams per liter. It is interesting to note that this source was so valuable that they "continued to pump from this bore to their newer Heriot Brewery in Roseburn into this century."[6] The Holyrood Brewery used wells of 20 feet (6 meters) and 110 feet (34 meters) until 1868, and wells of 145 feet (44 meters) and 119 feet (36 meters) until 1986. During this time, the wells were never contaminated.

Andrew Smith, who with his father became a partner of William Younger & Company in 1834, thereafter kept a notebook about brewing, and about other breweries. He observed that Jeffrey's Heriot Brewery in Grassmarket "bored deep 1856 and 1858—made a capital export ale ... one day in our brewery he attributed the keeping of their export ale—to the hard quality of their water—and no surface water being in their spring ... he said he bottled his strong ale two to three months old—and never had breakage."[16] Another brewery in Grassmarket had a well that gave 1,400 gallons per hour (53 hectoliters per hour) of 271 milligrams per liter total hardness, and 183 milligrams per liter of permanent hardness. W. J. Mitchell's 1951 notes on "Morningside Bore Water" give 309 milligrams per liter total dissolved solids, 220 milligrams per liter temporary hardness. The breakdown is 135 mg/L $CaCO_3$, 79 mg/L $MgCO_3$, 24 mg/L $MgSO_4$, 24 mg/L $MgCl_2$ and 8 mg/L NaCl.[16]

Russell Sharp states that Scottish "hard" water wells produced good hop- accented ales; "the water was hard, but not hard as in the sense of Burton water. Still, it makes good pale ales."[15] (Burton water measured 1,370 parts per million of calcium sulfate at one brewery, "with less magnesium sulfate.")[4]

"Hard" water has been pigeonholed as being of 900-1200 milligrams per liter of total dissolved solids, high in calcium and sulfate, with some sodium and chloride.[3] Hard Edinburgh water fits these limits nicely. A 1908 analysis of the water from a well at Bell's Brewery shows 921 mg/L TDS, composed of 269 mg/L $CaCO_3$, 116 mg/L $MgCO_3$, 190 mg/L $MgSO_4$, and 104 mg/L NaCl (hardness as $CaCO_3$, 500 mg/L).[16] Such water will yield excellent pale ales, but not quite of the Burton India Ale character. The chemist recommended addition of 500 to 570 mg/L $CaSO_4$ for "pale bitter ales" (increasing the hardness by 290-330 mg/L), and for "increased stability ... to add a further two ounces of gypsum to every barrel of wort in the copper [345 mg/L, 200 mg/L hardness as $CaCO_3$, for a sum total of about 1,000 mg/L hardness]. This is a procedure which ew [sic] have proved after long experience to give very clean worts, and it also increases the precipitation of those objectionable proteins which are one of the principle causes of instability."[16]

It was recommended that 40 percent of the gypsum be added after the mash, because "hard water will not have such free access to the malt as soft. The farina, therefore, will not be so effectually dissolved, and much of the saccharum [sugar], as well as of the flavor, will be left in the grains ... although it has been asserted by some brewers, that by using hard water the ale is prevented from fretting, and it is, at the same time, preserved."[4]

Robert Wallace of the Bass Crest Brewery in Alloa in 1873 listed "Substances of value for different classes of Beer" as being "Black Beers: Carbonates of Potash and Soda not less than five and not exceeding 40 grains per gallon [70-570 milligrams per liter], Pale Ales: Sulphates of lime, magnesia, potash and Soda. Carbonates of lime, magnesia, potash and Soda. Chlorides of Calcium, magnesium, potassium and Sodium. (The total mineral matter in the water should not be less than 30 grains per gallon [425 milligrams per liter])."[16] He also gave an analysis of five waters, three of which he considered appropriate for brewing particular styles of ale. He considered the second very good for mild ales, the third exceptional for pale ales, and both as promoting good keeping qualities. He states that the fourth provides "great character and weight" to black beers:[16]

	2	3	4
$CaSO_4$	288 mg/L	365 mg/L	21 mg/L
CaCl	-	-	-
$CaCO_3$	151	144	40
$MgSO_4$	-	644	-
$MgCl_2$	65	-	9
$MgCO_3$	51	-	5
$NaSO_4$	-	18	21
NaCl	4	52	123
$NaCO_3$	-	-	207
KSO_4	-	36	-
KCl -	-	4	-
KCO_3	-	-	36
(Hardness	439	727	62 as $CaCO_3$)

The 1870 well water analysis of an unidentified brewery (number 1 below) compared it to (2) "town liquor" and (3) the brewery's well water after treatment with 24 pounds

(10.9 kilograms) of sodium bisulphate and seven pounds (3.2 kilograms) of calcium chloride (presumably as dihydrate) per 200 barrels (235 hectoliters) of liquor (330 mg/L $NaHSO_4$, 97 mg/L CaCl):

	1.Well	2.Town	3.Treated Well Water
Total Hardness	300 mg/L	40 mg/L	[363 mg/L]
Alkalinity	240	30	40
Calcium	106	12	131
Sodium	22	7	-
Magnesium	12	-	-
Sulphates	90	10	356
Chlorides	42	15	87
pH	7.3	7.1	7.2

If town water were to be used for brewing, additions of 27 pounds of gypsum and 13 pounds of calcium chloride per 200 barrels of liquor were recommended.

Pollution of a well spelled disaster for a brewery. Ground-water contamination meant algae growth, off flavors and probable contamination in the brewery. Andrew Smith noted that "Melvins [Boroughloch] ... Aitchinson [South Back of Canongate] ... [were] not good, foxed."[16] "Fox" was the term for as-yet-unidentified algae and mold.

The urban Edinburgh breweries also faced "geological" pollution. "Problems had arisen with the water supply to the Craighwell Brewery in the early 20th century. Increased domestic and industrial usage resulted in the water table sinking, and pollution creeping in."[16] Dr. Brown of Scottish & Newcastle explains that "when water touches shale, it fouls—as happened in town. The water was very soft and pure; brewers had to add gypsum in the tuns. The water is now all contaminated. Breweries followed the fault out to

Slateford Road to get away from the polluted wells, as well as to avail themselves of rail sidings for shipment, and still have a steady supply of 51 degrees F [10.6 degrees C] water, without the shale contamination."[15] The shale contamination followed anyway. Water flowing through slate or shale leaches out iron and manganese and therefore is unfit for brewing. As the city extended, the aquifer was further depleted, so that by the mid 20th century, the 223 feet (68 meters) deep well at the Caledonian Brewery on Slateford Road was retired when the shale bank shifted there.

Andrew Smith, in an assessment of his competitor's water supplies, also records a disaster that struck several breweries in the late 1850s: "We always thought that the Abbey Brewery [at this time their main brewery] water most suitable for ales, as also Craigend ... Croft-an-Righ [that had been Robert Younger's St. Ann's, before Steel Coulson & Co.] we thought was much in the same strata ... we had put a well down ... but the gas came in and spoiled it." Furthermore, "Bernards ... water still continued to bubble in boiling. ... The Gas Works situated on the North Back of Canongate was held responsible for polluting the water supply in the vicinity."[16]

We have, then, a good record of the types of water used by Scottish brewers for their own ales, as well as the well-water and treatments used for brewing India ales for comparison. There is more to brewing beer than using the best-suited water, however. To quote Andrew Smith again, "Taylor, Anderson & Co. ... Mr. Thompson our brewer always spoke highly of their fine water and abundant supply, and he thought that if they only put in the malt they would be dangerous opponents."[16]

4

Malt

JOHN BARLEYCORN, A BALLAD

There were three kings into the East,
Three kings both great and high;
And they hae sworn a solemn oath
John Barleycorn should die.

They took a plough and plough'd
Put clods upon his head;
And they hae sworn a solemn oath
John Barleycorn was dead.

But the cheerful spring came kindly on,
And showers began to fall;
John Barleycorn got up again,
And sore surprised them all.

The sultry suns of summer came,
And he grew thick and strong,
His head weel arm'd wi' pointed spears,
That no one should him wrong.

The sober autumn enter'd mild,
When he grew wan and pale;
His bending joints and drooping head
Show'd he begun to fail.

His color sicken'd more and more,
He faded into age;
And his enemies began
To show their deadly rage.

They've taen a weapon, long and sharp,
And cut him by the knee;
Then tied him fast upon a cart,
Like a rogue for forgerie.

They laid him down upon his back,
And cudgell'd him full sore;
They hung him up before the storm,
And turned him o'er and o'er.

They filled up a darksome pit
With water to the brim;
They heaved in John Barleycorn
There let him sink or swim.

They laid him out upon the floor,
To work him farther woe;
And still, as signs of life appear'd,
They toss'd him to and fro.

They wasted, o'er a scorching flame
The marrow of his bones;
But a miller used him worst of all,
For he crushed him between two stones.

And they hae ta'en his very heart's blood,
And drank it round and round;
And still the more and more they drank,
Their joy did more abound.

John Barleycorn was a hero bold,
Of noble enterprise;
For if you do but taste his blood,
T'will make your courage rise.

T'will make a man forget his woe;
T'will heighten all his joy:
T'will make the widow's heart to sing,
Though the tear were in her eye.

Then let us toast John Barleycorn,
Each man a glass in hand;
And may his great posterity
Ne'er fail in old Scotland!

Rabbie Burns, 1787[2]

These well-known lines by Scotland's most renowned poet (and ale drinker) are obviously an account of the making of malt, from sowing the barley seed, to crushing it for the mash. As a farmer and then an excise officer at a time when ale taxes were levied upon the malt, Burns was undoubtedly well versed in the process of converting barley to the ale he consumed so prodigiously.[11]

In Burns' time, Scotland-grown barley was just emerging from obscurity on its way to becoming the most highly valued of barley for malt. At the start of the century Scottish malt sold for one-third the price of English malt, and was considered inferior to it until at least mid-century. "Two main types of barley were used. ... The first and most significant was common barley, generally grown in most of southern Scotland. ... The second was generally known in the southern Lowlands as 'bear,' and ... in the north as 'bigg' or 'big' ... the far hardier of the two ... Scottish grains were generally inferior to English. ... On these supplies the Scottish brewing industry primarily relied."[11] Bear and bigg were smaller than even common barley, were lighter in weight and had a heavier husk.[4]

The agricultural revolution of the early 18th century led to the sowing of better seed, on ground that was better cultivated. By the 1840s, Roberts would comment that "during the last half-century, Scotland has so ably cultivated barley, that she now supplies some of the finest in the market."[4] Ian Donnachie states that "East Lothian ... and the East Neuk of Fife grew the finest malting barley anywhere in Scotland. ... William Berwick ... said it was five percent better than any in Scotland and in good years a lot more. By 1805 it was "preferred by most brewers and distillers to English kinds."[11]

After 1820, Chevalier malt, developed by genetic breeding for uniform plumpness and a greater yield, would make up most of the lowland harvest.[3] Like most breweries, Andrew Smith's was using it in 1838, valuing it more highly than the English common barley that had commanded such a greater price a century before.[16] "The Chevalier barley has become a great favorite for malting, and justly so ... it is the best calculated for the maltster, as well as for the brewer."[11] Chevalier had a thinner husk, needed 20 percent less steeping time, and gave at least 2 percent more goods than did the older Scots barley. From the 1850s onward, Scots barley was unrivalled by imports. It was of low nitrogen, had a plump "pickle" (kernel), and gave excellent extracts. "From a bushel of the best malt weighing forty pounds, twenty-six—nay even thirty—pounds of fermentable matter, or saccharum, may be extracted [65 to 75 percent extract]; while from the same quantity of a different kind, twenty pounds are the most [50 percent]."[4] Robert B. Wallace's (Bass Crest, Alloa) notebook gives yields of 223 to 241 pounds per quarter of malt (336 pounds), roughly averaging 70 percent extract. "As a rule Well Malted Barley yields in available Extract 75 percent of its Weight in Malt ... 240 pounds per quarter."[16]

Such yields may seem exaggerated to present-day brewers, but the facts are that improved malting varieties have not so much increased yield to the brewer as they have to the farmer and the maltster. From a brewer's point of view, barley and malt quality have not gotten better, just cheaper. Roberts' accounts do reveal that not all malt gave such good extract; one brew yielded 61.4 percent extract, and he found extracts of 71 percent to be rare.[4]

Brewers preferred Scottish barley. My wife's uncle, Jack Horne, who worked for George Insill in Younger's maltings, expresses nothing but disdain for imported barley, conceding only that East Anglian was at least better than North African. Russell Sharp concurs, and believes that Scots malt is part of the flavor profile of Scots ales: "Scottish malts tend to be Scottish barley, which does taste different than English barley. It's there in the taste of the beer."[15]

Although Scottish brewers may have preferred Scottish malt, in fact they have used barley from around the world for over 150 years. The ships that sailed with Scotch ale on board returned to port with holds full of barley. Maclays bought barley from Oregon, California, Tripoli, Hungary, Moldavia, Smyrna and Cyprus as well as Scottish and English barley.[8] George Insill remembers that "prior to 1900, imported malting barley came mainly from New Zealand, Algeria, Smyrna, Morocco and the ... 'barley basket' of the

Danube region. Between the world wars roughly 40 percent of the malting barley was from the British Isles, the balance was largely European. In the 1950s, California came to dominate the barley market, and Chile and Australia became significant exporters. Barley was purchased if it met malting standards, without regard to origin. The imported barley was usually higher in nitrogen, averaging two percent. Before World War II, imported barley commonly gave 2.2 percent nitrogen. Scottish barley averaged 1.3 to 1.4 percent, so that a median nitrogen for the mix of malts going into the mash tun came to about 1.7 percent."[15]

Dr. Brown points out that the high-nitrogen barley had a great influence upon the acceptance of adjuncts, their heavy-handed use over the first half of the 20th century, and their continued acceptance today. George Insill concurs that "this was also the period of greatest adjunct use, 35 to 40 percent not being uncommon. Younger's switched from the sugars they had relied on previously to roughly 40 percent white maize, while McEwan's kept about 10 percent sugar in their formulations, with up to 30 percent white maize."[15] Sugar and saccharine syrups had been used in commercial Scottish brewing since at least the time of browsters peddling treacle ale, and were more heavily relied upon when economics and barley shortages dictated.

At least until the late 19th century, each of the Edinburgh breweries had their own maltings. Many breweries, and all of the larger ones, had several.[11] Control of malting gave them versatility. "Brewers in Scotland malt their own barley, which gives them the advantage [of malting it] into such malt as they judge suitable for their purposes."[4] Slack barley, or that which floated in the steep, could be roasted and made use of, so that operating their own maltings gave the brewers better economy as well as control of quality. "The maltster who manufactures for sale, is interested in allowing the acrospire to grow to a greater extent, than the maltster who makes for his own consumption ... because, by such means, he can obtain a greater increase [in size], and consequently, a greater profit ..."[4]

The malting season went from October to June, generally for 38 weeks a year. Many of the laborers at the maltings worked the fields in the summer, returning in late fall to begin drying barley for another season's maltings. Malting was hot, hard work; the laborers recourse was the "mawkie", a break for beer when work got too hot or too hard.[22] Each malting took "from 20 to 22 days [in England] ... and in Scotland 24 to 26 days."[4]

Barley had to be prepared for malting before it could be used. This process varied with its origin and condition. Barley that had been sun-dried to below 10 percent moisture content was the least troublesome, and gave the best percentage of germination. Younger's often bought barley fresh from the field at 20 to 25 percent moisture content and the kilns of the maltings under George Insill's direction spent up to seven weeks each season drying barley. It had to be dried gently, without exceeding 110 degrees F (43 degrees C), so that the germ wouldn't be killed. Barley from abroad was usually adequately dried already (to 12 percent), but came out of the ships' holds choked with CO_2, and had to be more vigorously aerated in the steeps.

Steeping was the next step in malting barley after drying. The barley and water in the steeps needed to be at a temperature of roughly 50 degrees F (10 degrees C) for the barley to absorb water evenly during the 60 to 75 hours that the kernels were soaked; well water was the ideal temperature. The water needed to be free from organic contamination as well, so that the malt wouldn't be spoiled. From the steeps the malt was laid into a "couch" for a 24 to 36 hours, to shed excess water, before it was "floored." The breweries all had traditional floor maltings; several of them continued in production into the 1960s. The "pieces" were floored at 10 to 16 inches (25 to 40 centimeters) deep, or less if temperatures were high. They were turned after 24 hours, when the malt began to "sweat." Temperatures were held at 50 degrees F (10 degrees C) by repeated turning. Malt would lie on the floor for eight to 14 days, depending on the barley type and the temperature, before it went to the kiln to be dried.[4] The Scottish malts were sprouted more slowly because they sprouted in a lower ambient temperature.

English malts today are germinated much as were Scottish malts in the last century. "Some allow the roots to get to seven-eighths of an inch (2.2 centimeters) long; others never wish to see them above half that length. ... The latter method seems preferable ... the process of malting is brought to a conclusion some time before the stem has ... burst the husk."[4] After germination, the malt was repeatedly turned to aerate it. This mellowed the malt before kilning.

A malt kiln was "a room, the floor of which is in general now laid with iron plates, perforated with small holes. ... Under the floor of this room is a moveable fireplace, in which a fire is made, when required, of coke, charcoal, culm, or wood."[4] Soon after 1840, double-floored kilns became common.

First the sprouting kernels went to the upper floor to be dried. The kernels were spread perfectly level to a depth of up to eight inches (20 centimeters) (the thinner, the better for quality, if not economy) and dried with a gentle heat, being turned four to five times a day. Next the temperature would be raised to not exceeding 158 degrees F (70 degrees C) to complete the drying. This took two and a half to three days. The malt was then was dropped to the lower floor. The greater heat at this level would cure the malt in about two days; the palest malts took only "one to one and a half

Caledonian's floor maltings. Photo by Greg Noonan.

days to cure."[11] "For high-protein malts, especially from imported barley, temperatures of 195 to 205 degrees F (91 to 96 degrees C) were necessary," George Insill reports, "to break down proteins. This was the rule from the mid-19th century until World War II. The extensive kilning caused sugars to amalgamate and gave a higher color. In the latter days of floor malting, higher temperatures and shorter times prevailed. For paler beers, the malt was very thoroughly dried at lower temperatures to prevent color formation, then it was finished off at 193 to 195 degrees F (89 to 91 degrees C) to below two percent moisture content."[15]

Pale malt only came within the technological abilities of the maltsters when coke became widely available in the late 17th century. "Coke is reckoned, by most, to exceed all others [Welch coal, straw, wood, peat, fern] for making malt of the finest flavor, and of a pale color, because it sends forth no smoke to affect it ..."[11, 23]

Most kilns did not employ indirect heating, which gave the palest and cleanest-flavored malts, until at least the 18th century. Very dark-roasted malts were also beyond the capabilities of the wooden kilns because of the intense heat required, and could not be turned by hand on iron floors quickly enough to prevent them fusing into a burned mass. The palest malts possible were no lighter than amber until the late 18th century, and the darkest no darker than brown until the cylindrical drum roaster appeared in 1817.[3] Even after that, amber and brown malts remained common, although "it was the universal practice of the brewers in Scotland to use only the palest malts for their finer ales."[4] Pale Scots malt then was darker than English pale is now; Roger Martin of Baird's Station Maltings in Essex gives 6 to 8 °EBC (3 to 4 °SRM) for Scottish pale, as opposed to 4 to 6 °EBC (2 to 3 °SRM) for today's English pale.

Roberts recorded that "there are several kinds of malt used by brewers; viz, the pale, the coloured, crystallized brown and black—the last three named are chiefly used by brewers of porter and stout."[4] By the late 1830s, brewers and maltsters were aware that paler malts gave better extract than high-dried malts and theorized that more of the starch was reduced to "saccharum." "Pale malt does not require the finishing-off heat to exceed 140 degrees F [60 degrees C], whereas that of the amber would require it to be 10 or 15 degrees higher, if the heat be judiciously applied; brown again still higher by 15 to 20 degrees F [8 to 11 degrees C], while to that of the black malt, the heat must be raised to such a degree as will bring it to the colour required. ... The best method of preparing them [brown and black] is with a very slow fire, until they are about two-thirds dry; the kiln is then shifted, and the malt allowed to remain for several hours in a heap to mellow ... and on the following morning should be returned to it. The depth should not exceed two

Russell Sharp with the 1869 copper in the Caledonian Brewery, Edinburgh. Photo courtesy of Merchant Du Vin.

inches. The fire at this time should be very brisk, and made of wood. The desired object is to snap and char the grain, and this the brisk fire will effect."[4] Brown malt was often called "snap" malt; because the kernels were exploded, it gave up its starch very readily.

Until the 1850s, slack and questionable barley were roasted without being malted. I surmise that the part that roast barley plays in the flavor of Scottish ales may have as much to do with legendary Scottish frugality as it does with any other design. However that may be, roast barley flavor is one of the important qualities of Scottish ales. It seems to gives them the same elusive, malty character that decoction mashing of undermodified malt brings out in continental lagers.

Although peat undoubtedly was used at times as a fuel, only one brewery in memory brewed a beer with a peat "reek" (the smoky flavor, valued in whiskeys). Russell Sharp and Charles McMaster fondly remember Edinburgh's Maclachlan's pale brew that was available into the 1960s. Mr. Sharp recalls its high final gravity and low hop rate, and that its light color belied its rich flavor. Like peated distilling malt, the malt for this beer was kilned slowly over a low heat, to a light color. Paradoxically, it was the largest-selling beer in Glasgow, where mild lagers outsell fuller-flavored ales!

George Insill pointed out the changes in malting over recent years. The floor maltings are gone; Scottish & Newcastle's Slateford Maltings, closed in 1986, was the last of the large-scale traditional floor maltings in Scotland. Its six malting floors, yielding 60 quarters per floor per week, with a maximum yield of 4,060 metric tons per year, sit idle and decaying. Since the energy crisis of the early 1970s, malting and brewing practices have been changed in the name of fuel conservation. Malts are presently dried to only 3 to 5 percent moisture content, and worts are boiled for a shorter time, which consequently gives wort of a paler color.

Nevertheless, Scottish ales remain among the world's most definitively malty beers.

> Then let us toast John Barleycorn,
> Each man a glass in hand;
> And may his great posterity
> Ne'er fail in old Scotland!

5

Hops and Bittering

Malt, water and yeast have always been basic and necessary components of beer. Bittering substances, to counterbalance the sweetness of malt, have a briefer history. Hops, which we tend to think of as the fourth necessity in beer, are actually only one of the most recent bittering substances to be used in brewing.

Hops were unknown in Britain until 1584, when Flemish immigrants settled in the southern English county of Kent. They used hops to flavor their beer in the low country, and brought the vine with them. Hops were not quickly accepted by Britons; numerous contemporary accounts vilify the "foreign weed." For a time after their introduction, beer brewed with hops was called "beer," to distinguish it from unhopped ale. Hops were even later gaining acceptance in Scotland than in England. They could not be cultivated within the short, cool growing season so far north. Consequently, hops were expensive, and moreover were "English," an attribute not likely to endear them to the Scots during this period of animosity between the nominally-united kingdoms. Hops did not become widely accepted in Scotland until the early 18th century.

Scottish ales have never been remarkable for their bitterness. Nevertheless bittering substances were used to flavor Scots ales long before the hop came to England's shores. Sweetness was balanced by herbs and spices often mixed in secret recipes called "groot," or grout. Most of the flavorings, especially before the period of colonial trade, were from native flora. The following is a list of bittering substances that Scottish brewers are known to have used:

Bog myrtle (*Myrica gale*) was accounted to cause rapid drunkeness.[17]

Broom, also known as gorse or whim, an intensely bitter common shrub, was boiled with the wort. Broom was the most important of the medieval Scottish bittering agents.

Dandelion roots and orange peel, in a warmed spiced ale that was popular in Glasgow up to the early 19th century.[5]

Darnel, a weed that grows amidst wheat, was commonly used in medieval Scottish brewing, and was reputed to increase intoxication.[17]

Ginger was used after the mid 18th century, in combination with a certain culture of a yeast and a lactobacillus which was called a "fungus." Ginger is a mild stimulant.

Juniper, also called gorse, or furze, is a spiny evergreen, the berries of which give the warm and pungent flavor usually associated with gin and schnapps. Juniper berries were added late in the boil, or hung in the ferment.[5]

Licorice, or its extract Spanish licorice has a sweet, astringent taste which increases thirst the more one drinks.[17] In the Highlands, the roots of wild licorice, called "carmeal," were used to flavor "cairm," a popular ale.[5]

Serviceberries, the fruit of a common tree (*Sorbus sp.*) were used.

Spruce shoots were very commonly used, and reputedly prevented scurvy.[5]

Watercress, like spruce, was used to prevent scurvy.

Wormwood (*Artemisia absithium*), a bitter and slightly aromatic narcotic intoxicant, was hung in the fermenter for up to a year. Broombush (wild wormwood) was also used in the Highlands.[17]

Many other herbs and spices were also employed in brewing before hops were begrudgingly accepted in the 1700s. The bittering agent that survived the longest in commercial brewing was quassia. It was the most important hops substitute of the 19th and 20th centuries. Quassia is the intensely bitter chips of dried bark from the South American *Quassia* and *Simaruba* trees (esp. *Quassia amora*). One pound of quassia could be substituted for 12 pounds of hops.

Scots never took to hops the way that the English did. Even in 1834, Andrew Smith reported a 140 shilling ale being returned to the brewery because, presumably, the hops in it were too aromatic and too fresh. "Use boiled hops for the present use and keeping—they injure the flavor a little but it will not pass with green hops. The first brewing we made October 30th was nearly all returned having been hopped with Green."[16] John Carter's 140 shilling keeping ale of 1838 used spent hops for bittering. It was common practice for Scottish brewers to boil their second worts with the hops left from the first.[16]

Six varieties of hops were commonly used before Goldings, the first commercially selected variety, was introduced in 1785. The most important of these six was Farnham Pale, a hop that probably had very little bitterness.[3] Another variety, the North Clay hop, was distinguished for its "pre-eminence in rankness." North Clay was "grown on the stiff clays of Nottinghamshire ... best adapted for strong, keeping beers."[4] Kent Goldings, first associated with porters, were considered by Roberts to be of "preeminent distinction ... considered as uniting flavor with strength," so much so that by 1840, "in Edinburgh ... nine-tenths of the hops which were used in brewing are grown in the county of Kent."[4] Bavarian and Belgian hops were used almost interchangeably with Farnham and Goldings by Scots brewers.[3] Fuggles became almost as popular as Goldings with Scots brewers after they became commercially available in 1875.[3]

The principal reason for Scotch ale's historic success in export markets was its superior keeping quality. Lower gravity beers had poor keeping qualities, whereas the alcohol in strong beers preserved them. Presumably, English barley wines and strong ales were not as marketable, perhaps because oxidation of the fusel alcohols produced during high temperature fermentations gave negative flavor changes as these ales aged. Not until the keeping quality of India ale indicated that hops had a preservative nature, did hops come to be appreciated. Though far lower in alcohol content than strong Scotch ale, India ale proved to travel and store even better. "Even keeping beers for the home consumption, were they made from such low gravities as some [India ales], would certainly not stand over the summer. ... India pale beer being so highly impregnated with the finest hops, has not only been appreciated in India as a refreshing beverage [but also elsewhere]."[4]

We have a fair amount of evidence about the amount of hops used in 19th century beers. Unfortunately, without knowing the alpha acidity of the hops, their percentage of utilization, or the bitterness of the ale against a standard scale (like the International Bittering Units, or IBU), knowing only the amount of hops used in a particular brew is of little practical use.

Modern homebrewers use Homebrew Bittering Units (Dave Line's Alpha Acid Units, renamed HBUs) to define the hop bitterness in recipes. HBUs are the measure of the alpha acidity of a particular quantity of a hop. One HBU equals one avoirdupois ounce of hops of one percent alpha acid. Using HBUs to define hop additions gives reasonable accuracy and repeatability; the method does not account for oxidation loss of alpha acids since the hops were last analyzed for alpha acid content, or for actual utilization of the hops in the kettle, which can range from 10 to 30 percent.

Commercial brewers measure the bitterness in the finished beer to avoid inaccuracy introduced by these variables. Two standards are commonly used. International Bittering Units (IBUs) measure isomerized alpha acids; Bitterness Units (BUs) measure the total bittering substances, which includes oxidized beta acids. The latter method is used by the American Society of Brewing Chemists. It is more accurate because it measures the broader spectrum of bittering substances, and thus permits more exact repetition from brew to brew.

These tools weren't available to brewers until well into the 20th century. Previously, the most common method of recording bitterness in a recipe was to state the quantity of a particular hop variety used in a given volume of wort. This method took no account of the amount of malt used,which greatly affects the perception of bitterness. Scottish brewers, on the other hand, expressed hops not in relation to the volume of wort, but rather in proportion to the amount of malt used. Hop rates were expressed as pounds of hops per quarter of malt. A quarter is 336 pounds.[3] As sweetness increased with a larger malt charge, bitterness was increased proportionally. This method gave as good a measure as was possible at the time, but it could not account for variances in the alpha acidity of the hops. To adjust for bitterness lost during hop storage, 19th-century brewers increased their hop charges progressively for brews later in the year. However, the method is so inaccurate that it is no longer used, and even homebrewers using HBUs have far greater control over their bitterness levels than 19th-century brewers had. Therefore, recipes, records and ledgers of the period are only useful as indicators of the bitterness of the beer that was brewed if we estimate the alpha acidity of the hops of the day and hop utilization in the kettle.

It is nearly universally assumed that British hops of the 19th century ranged between two and five percent alpha acid. Utilization in the 19th century cannot have differed much from the present day range of 15 to 30 percent. The rate of utilization depends on the boil time and wort density.

With these assumptions, we can read Roberts' formulas. He writes, "For ales of high gravity, say 105 to 115, the quantity of hops ranges from four to ten pounds per quarter; in the winter brewings the average being about six pounds, and in the spring brewings, for the ales intended for summer consumption, about eight pounds."[4] These proportions translate as:

For winter brewing use 0.29 ounces of hops per pound of malt (18 grams per kilogram), or 0.92 ounces of hops per U.S. gallon of wort (6.75 grams per liter).

For spring brewing use 0.38 ounces of hops per pound of malt (24 grams per kilogram), or 1.22 ounces of hops per U.S. gallon of wort (nine grams per liter).

Robert continues, "The brewers of Scotland use for each quarter of malt ... from four to eight pounds, according to the quality of the ale, and the season of the year. In winter brewings, six pounds of hops for the best ale, and four for the inferior kinds, may be considered a fair estimate. Our practice of brewing, from January to March, was to allow ten pounds of hops per quarter of malt, when the wort was 95 to 100 of specific gravity. ... The hops we preferred were the East Kent, and Worcestershire."[4]

At roughly the same time as Roberts was writing, Andrew Smith observed that John Carter used 220 pounds of Farnham hops for 60 barrels of 140 shilling keeping ale.[16] In 1873, Robert Wallace of the Bass Crest Brewery recorded in his notebook that 336 pounds of hops were used for 105 3/4 U.K. barrels of 72 shilling and 84 shilling pale ale (2.28 pounds per U.S. barrel [0.88 kilograms per hectoliter]), and 322 pounds for 103 3/4 barrels of 72 shilling pale ale (2.23 pounds per U.S. barrel [0.86 kilograms per hectoliter]).[16]

These and other records of hop rates give us some indication of bitterness. For Wallace's brews of about 1.068 SG (16.4 °B), if we assume that kettle utilization was 25 percent, then the hops used would have been three percent alpha acid, and his ale rated about 66 BU. This figure is not unreasonable in that Scottish pale ale was reported to be less bitter than that of Burton or London. Assuming that Roberts' "better" ales were 1.070 SG (17 °B), then six pounds of hops per quarter of malt (1.75 pounds per U.S. barrel [0.93 kilograms per hectoliter]) at 25 percent utilization would have required hops of three percent alpha acid as well, and produced a bitterness of about 66 BU. A strong ale of 1.095 to 1.100 SG (23 to 26 °B) would have poorer hop utilization, perhaps 20 percent. Ten pounds per quarter (2.91 pounds per U.S. barrel [1.6 kilograms per hectoliter] of 3 percent alpha acid hops may have produced even less than 60 BU to those strong brews. Ultimately this formulation remains conjecture, but with a good guess at the alpha acidity of

19th century hops, and the rate of utilization, we can read from these records an approximation of the bitterness of the old ales.

We do know that fresh hop aromatics were considered a liability rather than an asset. For instance, Andrew Smith notes about Jeffrey's of Edinburgh that they "made a capital export ale ... no green hop." Furthermore, we know that hops were not well stored: "Hops ... deteriorate by age to such an extent, that at the expiry of one year they become less valuable by twenty-five to thirty percent."[16] Although tightly compressed modern Goldings stored cold will not lose more than 15 percent of their bittering value over 12 months, packaging, transport and storage conditions of the early 19th century caused quicker oxidation and loss of the soft resins and essential oils. One modern study indicates that the "noble aroma" associated with continental lagers is a product of oxidation after a moderate period of hop aging. Goldings-type hops share this characteristic with classic Bavarian varieties.[24] This observation allows us to infer that hop oxidation products may have been part of the character of 19th century Scottish ales.

Comparing hop rates of the period for Scotch and India ales shows that the latter used twice as much hops as the former; India ale at OG 1.063 (16 °B) used 1.7 ounces per U.S. gallon (12 grams per liter) in the kettle, and was dry-hopped with one-fourth ounce per gallon (2 grams per liter). "The Burton brewers average from about 20 to 22 pounds per quarter, and they generally prefer the East Kents; but the average for others, for this beer [India ale], is about 16 to 18 pounds. ... Six pounds per quarter are first put into the copper, and boiled for twenty minutes, after which eight pounds per quarter are added, and with the first quantity boiled for fifty minutes longer ... and leaving the remaining eight pounds per quarter for the second wort."[4] By weight,

Burton brewers used 0.66 ounces of Kent hops per pound of malt (41 grams per kilogram, 1.3 ounces per U.S. gallon of wort at 1.063 OG), and elsewhere brewers used 0.44 ounces per pound of malt (128 grams per kilogram, 0.9 ounces per U.S. gallon for wort at 1.063 OG).

Bitterness and hoppiness were not dominant factors in Scotch and Scottish ales of the last century, nor are they today. Hops are primarily used to provide a balance to the malt sweetness. Scottish & Newcastle's Dr. Brown summarizes the current Scottish attitude regarding hops, saying that "hops are used for bitterness in Scots' brewing, and not for flavor. Whether they be Fuggles, Goldings or an English bittering hop, at 25 to 26 IBU, hop bitterness is subdued and there is no real hop flavor."[15]

6

Yeast

Scotch ale and Scottish ales share a host of common characteristics, a clean maltiness being the most obvious. Part of this is due to the character of Scottish brewing yeast. These yeast don't exhibit overt fruitiness or other estery characteristics and therefore allow the malt flavor to come through cleanly. One reason, of course, is that these *saccharomyces* strains are able to ferment at lower temperatures than many other ale yeast strains. This low temperature viability is one of the primary characteristics of Scottish ale yeast. Although Scottish brewing yeast produce low levels of esters and fusel alcohols, the ales of several contemporary breweries which ferment at higher temperatures than was the rule in previous centuries have a distinctive diacetyl aroma.

Another major trait of these yeast were that they were relatively under-attenuative. The ales historically brewed with them did not ferment down to quarter gravity; even many current brands are only attenuated to one-third of their original gravity. We know that they were relatively flocculant strains because the brewers regularly "beat" the head back into the beer to force it to ferment out.

A modern Scottish strain that we have used at The Vermont Pub and Brewery of Burlington forms short chains. Consequently it is not a "dusty" strain, but it does not form long chains and flocculate excessively, either. It will also ferment at temperatures between 50 and 60 degrees F (10 and 15 degrees C), and is very neutral in character. This strain completely ferments maltotriose, but historical strains may not have. Until the late 1800s, even the most sophisticated breweries did not have the technology available to attemper fermentation artificially. In current practice, breweries use refrigeration to lower the temperature of the beer rapidly when they reach their target final gravity. The sudden temperature drop shocks the yeast, and it drops out of suspension. In the last century, only high mash temperatures combined with under-attenuative strains could have accomplished this rapid precipitation.

A third shared distinction of these strains is that they are alcohol tolerant. Today, Edinburgh ales and wee heavies are often of eight percent alcohol by volume (6.4 percent alcohol by weight). In the last century, brews of eight to ten percent alcohol by volume were usual. Scotch ales of this strength may never have fermented past one-third gravity, but from an OG of 1.120 (28 °B), the beer would have contained over 9 percent alcohol by volume (7.2 percent alcohol by weight) by the time it had fermented down to only 1.050 (12.3 °B). The yeast had to be able to repeat this feat of fermentation brew after brew without being poisoned by the alcohol. At least one modern strain is use is capable of yielding 10 percent alcohol by volume (8 percent alcohol by weight).

These three factors are probably what selectively determined the strains used by Scottish brewers.

Roberts is an invaluable source on yeast, providing detailed information about Scottish practice before the scientific revolution of the late 1800s. He advises against the "greater evil, namely, that of over-storing [the] tun," and recommends making a starter at a higher-than-pitching temperature to activate the yeast, and to judge the yeast's viability by the strength of the starter. Roberts may not have known that yeast were living organisms, but he did know how important they were to a brew. "Yeast are technically called 'store.' ... Desired flavour and attenuation ... depends upon ... stock which has been of equal gravity to the wort for which it is now required. Should the ferment, or store, be the product of unsound worts, or be in any respect tainted, it will impart its noxious quality to the worts with which it is combined. ... Should the brewer ... make use of store of an inferior quality ... nine times out of ten the ale will prove turbid and 'yeast bitten.' ... The store which ought to be preferred is that produced from the last stage of vinous fermentation, namely, from the stillions [the casks on their racks before bunging]. Care should be taken that it has not been long exposed to the atmosphere, or allowed to remain long in a warm situation; because, in either case, it will fret or ferment, and consequently be deteriorated in stamina."[4] Dr. Brown adds to this, noting that "Yeast has a good memory. It will subtly change within a few generations if it is subjected to change. Brewers then, as now, had to really cater to the needs of their yeast."[15]

Surprisingly, most breweries did not maintain a house strain, and believed that using yeast from another brewery improved the vigor of their ferments. Although Roberts quotes Black as saying, "I have worked in a large establishment from year's end to year's end, without ever having the least occasion for a change of yeast, and could do so again at any time, and my fermentations shall be as healthy and vigorous as any one's,"[4] this was not the usual practice. Younger's records of the mid-1800s show yeast obtained from Dryborough, Aitchinson, Blair, Campbell and others, apparently randomly chosen on the basis of which brewery had healthy yeast to spare on that brewing day.[16] The conclusion is that the Edinburgh brewers, if not brewers from the neighboring towns as well, shared a common mixed strain.

Roberts argued that borrowed cultures were more likely to produce consistent results than a house strain, indicating that diverse conditions were required to prevent any component strain of the mix from dominating.[4] Today single-cell cultures are used, and at least three breweries use the same *saccharomyces* strain.

Any *Cerivisae* strain that is neutral, ferments well at below 65 degrees F (18 degrees C), and gives no more than 65 to 70 percent apparent attenuation should be appropriate for brewing Scottish ales; alcohol tolerance is a further requirement for brewing the stronger Scotch ale.

Brewers also recognized that the volume of yeast to be pitched was important. "It is generally affirmed, that the higher the gravity, and the lower the heat at which the fermentation is commenced, the more yeast will be required to obtain a uniform attenuation. ... When the temperature averages about 42 degrees [5.5 degrees C], and the gravity of the wort ranges from 90 to 120 [21.5 to 28 °B], two pounds and a half of yeast are commonly used for each barrel of wort (0.97 kilogram per hectoliter)," based upon a gallon of yeast weighing 11 to 12 pounds.[4] Because Roberts gives not only the pitching rate by weight, but the weight to volume ratio of the yeast slurry as well, we can estimate the cells per milliliter pitching rate with confidence. His figures give 0.644 fluid ounces of yeast weighing 0.925 ounces per gallon of wort (U.S. measures). Duplicating this with a yeast strain currently used by a Scottish brewery gives a paste the thickness of putty, and a pitching rate of 18 to 22 x 10^6 cells per milliliter. For homebrewers things are not so well defined; the best that can be done is to pitch more yeast than would be used for a lager. For five gallons, one to four quarts of a very active starter will be required. The pitching rate is as important a factor in creating Scotch and Scottish ales as the malt, the mash temperature and the hop rate.

7

Scottish Brewing

"March 1, 1836. £.5 Ale—Malt, 20 Quarters; Hops, 160 lbs.—Temperature 42 degrees.

"Commenced brewing at 4 A.M., by turning into the mash-tun thirty-two barrels of liquor at 200 degrees. When reduced in temperature to 180 degrees, shot twenty quarters of pale malt into mash-tun; and raked and mashed with oars forty-five minutes. Finished mashing at 5 ho. 15 min., and strewed a bushel of grist over mash and covered up. At 8 o'clock set tap, uncovered, and commenced sparging at the same time, with liquor at 190 degrees, and continued the operation until thirty-two barrels were sparged. Wort running quite fine with good appearance; temperature 148 degrees, and 110 gravity. At 11 ho. 30 min., ale wort all in copper, which gauged forty-eight barrels, at gravity 83, Allan's Saccharometer. Previously, however, shut tap, and sparged on mash fifteen barrels of liquor for table beer.

"Weighed one hundred and sixty pounds of the best East Kent hops, and put them to the wort in the copper; and 12 wort came through, boiled briskly one hour and twenty-five minutes, and at 1 ho. 30 min. cast copper. At 2 ho. 30 min. spread in coolers. At 10 o'clock p.m. pitched tun with eight gallons and a half of store weighing 90 lbs., and let down wort at 50

degrees, which gauged in the tun thirty-six barrels and a half, gravity 103.5.

"The following table shows the heat of the tun during fermentation:

March 2,	*1 o'clock P.M.*	*50°*
March 3,	*9 o'clock A.M.*	*0°*
March 4,	*9 o'clock A.M.*	*52°*
March 5,	*8 o'clock A.M.*	*56°*
March 6,	*10 o'clock A.M.*	*58°*
March 7,	*8 o'clock A.M.*	*58.5°*
March 8,	*8 o'clock A.M.*	*59°*
March 9,	*1 o'clock P.M.*	*60°*
March 10,		*61°*
March 11,		*62.5°*
March 13,		*63°*
March 14,		*63 °*
March 15,		*62°*

"March 15—Removed the gyle from the tun into the square, the temperature of which when let down was 62 degrees, and gravity 43.

March 16, 10 A.M.—Cleansed into hogsheads and barrels.

The summary of the foregoing example is as follows:
Ale, 36 barrels, at gravity 103.5 = 3726
Beer, 10 barrels, at gravity 40.0 = 400
Quarters of malt used, 20) 4126
Value extracted from each quarter: 206 3/10"

The Scottish Ale Brewer
and Practical Maltster, 1847

"Let me Exhort my Countrymen to brew their Ale from the Softest Water, the Palest Malt and the Most Fragrant Hops." George Younger, 1779.

THE MASKIN' LOOM

At the time that Robert's wrote *The Scottish Ale Brewer and Practical Malster,* kiln technology had made light colored malts available to the brewer, roller mills had nearly replaced millstones for crushing malt and false bottoms were fitted in most Scottish mash tuns. (Even so, wooden mash tuns, or "maskin looms" were still common in 1890).[11] Brewhouses held three coppers, one for boiling the wort, one for mash liquor and the third for sparge water. The brew day began with the heating of the mash water; "the heats generally employed by brewers in Scotland ... are considerably higher than those of English brewers. Those heats range from 178 degrees to 188 degrees [81 to 87 degrees C], and even to 190 degrees [88 degrees C]."[4] The heats referred to are those of the mash and sparge liquors; the mash liquor went into the tun to preheat (rather than being introduced from below after the malt was in the tun, as was the English practice).

Ambient and malt temperatures were usually 45 degrees F (7 degrees C) or less, so that liquor temperatures even as high as 180 degrees F (82 degrees C) would give a mash temperature of no more than 158 degrees F (70 degrees C). Mash thickness in the 19th century was not remarkably different from that currently recommended for infusion mashes. "The quantity of liquor generally used for mashing is about one barrel and three firkins per quarter [29 fluid ounces per pound, 2 liters per kilogram], and sometimes only one barrel and a half [25 fluid ounces per pound, 1.7

liters per kilogram]."[4] Robert's notes about brewing India ale are an indication of the temperature of the mash for Scotch ale. He states that at ambient temperature of 40 to 45 degrees F (4 to 7 degrees C), when the malt for India ale was stirred into liquor at 168 to 170 degrees F (76 to 78 degrees C) it gave a mash temperature of "rather less than 150 degrees [65.6 degrees C]."[4] Water at 180 degrees F (82 degrees C) would be expected to give a mash temperature of 158 degrees F (70 degrees C), and less if the ambient temperature were lower. Unfortunately, the limitations of period thermometers meant that 19th century Scots brewers did not record the actual temperatures in the mash itself; records of runoff temperatures ("1st Falling Heat") that range between 149 and 152 degrees F (between 65 and 76 degrees C)[16] indicate mash temperatures of around 156 degrees F (69 degrees C), for a mash resting two hours.

The malt was mixed into the water with oars. Balled starch had to be broken up; the Scottish method surely produced more balling than if the wet had been added to the dry. The old Scottish method of mashing in survived until after 1853, when Scotsman James Steel patented his Steel's masher. Combining the grist and water together inside a helical screw, Steel's invention caused the grain to be more thoroughly soaked than laborious hand mixing ever could.

After 45 minutes or more of mashing in, the mash would rest for two to three hours to saccharify. Because the mash began at 158 degrees F (70 degrees C) or so, beta-amylase activity was minimal. Although the mash time was long, by the time that the temperature dropped to a low enough point to be optimal for beta-amylase activity, the enzyme would have been largely inactivated by the high initial temperature. Some alpha-amylase, on the other hand, would have remained active for most of the mash period, having been thermally protected by the thickness of the mash.[25]

Once the mash was finished, the wort was slacked into the underback and sparging was begun when a fifth of the wort had been run out, if it was not begun immediately.[11] A volume of sparge water equal to, or up to 40 percent greater than the quantity used for mashing in, was sprayed onto the mash from a ring encircling the top of the tun. The mash was kept just covered by the sparge water, as is still usual practice today.[4] Thus a depth of water does not rise above the settling husks, which by its weight could unnecessarily compress the filter bed. Sparge temperatures were higher than modern research would recommend; in the 1800s the Scots regularly sparged with water at 190 degrees F (88 degrees C), though 172 degrees F (78 degrees C) is now thought to be the optimal temperature for sparge water.[4] Given that the ambient temperature was so low, and the temperature in the mash dropped radically during the prolonged mash, the temperature within the mash probably never exceeded 175 degrees F (79 degrees C), and probably seldom reached that temperature. Again, temperatures taken in the underback indicate this; mid 19th century records of "3rd Falling Heats" from William Younger's brewery seldom exceed 160 degrees F (71 degrees C), suggesting that internal mash temperatures might not have exceeded 165 degrees F (74 degrees C) during sparging.[16]

Roberts states that "the great aim of all brewers is, that the wort, when it flows from the mash tun, shall range from 147 degrees to 152 degrees [64 to 67 degrees C]."[4] Records bear out that this was generally the case, and 147 degrees F (64 degrees C) or below was at least as usual as 152 degrees F (67 degrees C).[4,16] The correct mash and sparge temperatures gave a transparent wort with a "fine, light pearly head of considerable height. If on the contrary, the colour be deeper than that of the malt employed, and if the froth be of a reddish, fiery appearance, and deficient in height, although it be transparent, there is every reason to fear that the mashing heat has been too high." If the wort had a "turbid, dead appearance" (i.e. starch cloudiness), then the mash temperature was too low.[4]

Roberts quotes Mr. Booth's description of English mashing from his *Art of Brewing* to compare it with Scottish practice. English strong ale was "double mashed," first with water at 180 degrees F (82 degrees C). Then the wort was run off after a time, and the grist was mashed again with water at 185 degrees F (85 degrees C). Table beer would be made from a third mash, using water at 190 degrees F (88 degrees C). The increasing temperatures would have leached out unconverted starch from the malt, so that conversion by alpha-amylase would have continued during the second and third mashes.

Roberts defended the Scottish practice of sparging against the English double mashing method. "The process of sparging is, in my opinion, decidedly preferable to a second mash for ale worts, and has ever been considered in this light by the whole of Scottish brewers."[4] Presumably, the advantages he means are a clearer wort, with less extraction of husk polyphenols, silicates and fatty acids, and therefore beers with better head retention. Modern Scots' preference for a pint with a head may very well reflect Scottish traditional sparging practice; English ale drinkers have always seemed less concerned with a topping of froth.

On the other hand, Scots brewers sparged with water at a higher temperature than is usual today, using liquor of up to 190 degrees F (158 degrees C).[4] This would necessarily have caused extraction of poorly converted starch, as well as extraction of husk phenols. The only remedy to both of these problems is a long period of fermentation and cellaring.

During conditioning, hydrolysis slowly breaks down intermediate starches. Polyphenols and proteins not scavenged by the protein-and-mineral-starved yeast precipitate out, thereby enhancing the beer's stability.[25]

Like the English, Scots brewers generally ran the later weak wort runoff into a separate kettle, to be boiled and fermented for table beer. The tails of the sparging were relegated to the production of twopenny ale, which contemporary detractors assure us was peaty, harsh and fuellike.[11] The descriptions of twopenny are of a brew high in husk phenols and clouded with starch.[4,11] These assuredly made it a loathsome brew, but these phenolic compounds and starch fragments just as assuredly supported the yeast in the long cellaring of Scotch ale.

THE COPPER

Worts in the copper were not treated much differently by the Scots than they are by modern brewers. "The time of boiling for ale wort ranges from one hour to one hour and a half. ... The boiling of this wort for a longer time than one hour extracts the coarse flavor of the hop, while the fine aroma, being more evanescent flies off with the vapor ... however ... a sufficient time for boiling is necessary, not only to extract the aromatic flavour and the preservative principles of the hop, but to coagulate the super-abundant gluten of the first wort."[4]

Like the English, commercial brewers in Scotland commonly added sugar adjuncts in the copper. How much is a matter of debate, fueled by contradictory records from periods when malt was cheap and plentiful on the one hand, and from when it was scarce on the other. The records do clearly show that adjuncts were more liberally used in brewing India ale than they were for for Scotch ale.

Donnachie's research shows that as late as 1886 only 4 percent of total brewing ingredients were sugar. Edinburgh brewers used the least, and Glasgow brewers the most (10.5 percent).[11] This regional difference is reflected in the beer styles that each city brewed and consumed; after 1850 Glaswegians progressively drifted towards lighter and lighter beers.[5,12]

Another aspect of Scotch ales is the caramel character that they pick up in the copper. George Insill emphasizes that historically "color precursors from long kiln times would develop in the copper, especially when two-hour boils were commonly practiced to degrade proteins to give the ale better stability."[15] Even a present day 70 shilling heavy has a distinctive caramel flavor. Drinking a Scots ale, one cannot help but be struck by it; yet only one Scottish brewer uses crystal malt in their brews. The caramel flavor is from caramel sugars in some cases, but more often from

Direct fired coppers, built in 1869, are still in use at the Caledonian Brewery. Photo by Greg Noonan.

carmelization in the kettle. It is a maltier flavor than that obtained from crystal malts.

At the end of the boil, worts were allowed to settle only briefly before being run into the coolers. As pointed out previously, this practice is very different from English one, and the configuration of the hop-back was very different as well. "In England, the hop-back is a square vessel of wood or iron, of a capacity to hold more than the contents of the wort copper, with a false bottom perforated with small holes. ... The wort is allowed to remain in the hop-back for a short time before it is run off into the cooler; by this means the hops subside at the bottom; and the wort, when drawn off, filters through them, leaving its impurities behind it. In Scotland, the brewers ... use a square or oblong wooden box with a temporary bottom made of haircloth. ... The wort immediately disengages itself from the hops, carrying along with it most of its impurities. ... None of the grosser particles can run through it [the haircloth bottom], it will only allow a small portion of the fecula of the hop to escape; ... this fecula is highly beneficial, both as a preservative in the coolers, and afterwards as an agent in inducing vigorous fermentation."[4]

Roberts thought it unwise, to allow this "fecula" (trub) in the runoff, since the sediment gave a "disagreeable coarse mealy flavor", although he presents Robert Stein's view that it made fermentation more vigorous.[4] The trub (protein/phenol complexes), though it may have ruined the flavor of a young brew, would have sustained yeast metabolism and would have been hydrolized to a host of pleasing flavor compounds over the long storage period common for old Scotch ales.

The hop-backs and coolers that Scottish brewers used to employ have not become entirely obsolete; the brew-house at Traquair House uses them today. "Coolers, formerly used in Scotland, were, as in England, wooden floors of large dimensions, having sides not exceeding six to eight inches [15 to 20 centimeters] in height."[4] Wort was run into three to six of them consecutively at a depth of not over 1 1/2 inches (3.8 millimeters).

Hop-backs and open coolers subject the brews to the most danger of contamination. Until after 1860, brewers had little understanding of what contributed to contamination. Although the heat of the wort itself would have pretty well sterilized the coolers, mold and bacteria would have leached into the wort and given it off-flavors.

"The worts are very prone to ferment spontaneously in the coolers ... acetification takes place before they even enter the fermenting tun. The result of this spontaneous fermentation is, in the language of brewers, called foxed worts; a disease which is known by the worts producing on their surface mouldy spots of a reddish colour, and by their emitting a disagreeable odour ... an unpleasant flavour, and preventing it from becoming transparent. ... The fox was scarcely known in Scotland till lately; because the strong ale brewers confined their operations to the colder months of the year, the coppers being silent from May to October."[4] Presumably because of the proneness to foxing Roberts recommended that "strong ale worts should not remain in the coolers for more than nine hours, nor less than four."[4]

THE GYLE

Where Scottish brewing practice differs most from English, and where, as Tennent's David Johnstone emphasizes, it most nearly approaches lager brewing, is in fermentation. Although the Scots breweries did not generally ferment at less than 50 degrees F (10 degrees C), and ferments usually got up to 60 degrees F (16 degrees C) or higher, "gyle" (fermenting vessel) temperatures were cool enough that yeast activity and ester production was subdued. Their temperatures were not remarkably different from those used by modern lager brewers. Furthermore, Scots ales underwent prolonged conditioning in hogsheads at cold temperatures, which also is more like lager beer than like traditional English ale.

Even today, Scottish brewers ferment somewhat cooler than English brewers, using strains that do not produce many esters. Dr. Brown of Scottish & Newcastle advises that oxygenation of the wort must be minimal; "low oxygen and a low initial temperature are necessary to control esters by reducing yeast growth. Five-fold yeast growth is even too great for Scottish beers; the pitching rate must be great enough that the yeast needn't exceed three-fold growth, to control esters."[15]

Temperature control is very important for developing the unique character of Scotch ale. "Scottish fermentation techniques before the 1830s ... used temperatures in the range from 44 to 58 degrees [7 to 14 degrees C], and averaging 50 degrees [10 degrees C] ... [for] up to 21 days, the corresponding time in England being five or six days."[4] Donnachie and others who refer to temperatures in this range are referring to the temperature at pitching; brewing logs confirm that temperature of fermentation rose to 60 degrees F (16 degrees C) or higher before the yeast had depleted the extract.

Roberts described a typical fermentation. When "pitched at 50 degrees [10 degrees C], there is little appearance of motion during the first twelve hours. ... After the space just mentioned, however, the agitation becomes quite apparent, until, after about forty or fifty hours from the commencement, the whole surface is covered with a thick white curly foam. This foam continues to rise in height, until it assumes a rocky appearance, termed the cauliflower

Wort working in the gyles at the traditional Traquair House Brewery. Photo by Greg Noonan.

head. About this time it is customary to beat it in; an operation which is repeated morning and evening until the tun has increased from eight to ten degrees of heat [up to 58 to 60 degrees F, 14 to 16 degrees C]. ... The operation of skimming is seldom or never performed in Scotland, except when it becomes necessary to check by this means too vigorous a fermentation."[4]

At this period, when the usual Scottish pitching temperature was 50 degrees F (10 degrees C), a typical English brew began at 65 degrees F (18 degrees C). The flavor impact of the increased esters produced at the higher temperature distinctly separates English ales from those of Scotland. Scottish ales are far mellower, and evidence very little fruity, vinous or fusel character.

Fermentation temperatures began an upward trend in the 1830s. Scots were then "setting their tuns" at 53 degrees

F (11.6 degrees C), and rising up to 65 to 67 degrees F (18 to 19.5 degrees C) during fermentation. By mid-century, pitching at 55 degrees F (13 degrees C) was common, with temperatures rising to about 65 degrees F (18 degrees C) for ales of less than OG 1.090, and up to 62 degrees F (17 degrees C) for ales of OG 1.090 to 1.128 (21.5 to 31 °B).[16]

Contemporary commercial brewers rely on refrigeration to brew year round; in Scotland, as elsewhere in Europe, brewers had to restrict production to the colder months of the year to avoid both runaway fermentations and contamination. Donnachie quotes William Black's *Practical Treatise on Brewing* that, " 'Keeping beers' should be brewed in frosty, or at all events cool, open weather, which may be expected in December, January, February and March."[11] In Scotland brewers were more fortunate. The cool climate gave them a brewing season from November to April.

The difference between English and Scottish practice was well-examined by Roberts, writing at the very time that Scottish brewers were introducing India ales to their product lines. The Scots brewers had to learn how to amend their techniques to produce the characteristics of India ales. Roberts summarized the essential differences in a description of English ale brewing. Pitching rates were 20 percent lower for the lighter brew. The English began fermentation at up to 65 degrees F (18 degrees C), allowing the temperature to rise to 67 to 72 degrees F (19 to 22 degrees C). Primary fermentation might be completed within as little as 24 to 30 hours, and the beer was transferred to casks for secondary fermentation. The ale was "well-roused and cleansed into hogsheads or puncheons, and ... further attenuated down from 20 to 24 Allen [i.e. 1.020 to 1.024, 5 to 6 °B]." Pale ales, with starting gravities of 1.050 to 1.070 (12 to 17 °B) fermented out to FG 1.005 to 1.012 (1.3 to 3 °B) after 14 to 20 days of secondary fermentation in the puncheons. Finally, the ale was racked to hogsheads, dry-hopped at a rate of one-half pound or more per U.S. barrel, and bunged down for final conditioning.[4]

Perhaps the example of India ale brewing influenced Scottish brewers to increase fermentation temperatures for traditional ales. The lighter-colored, lighter-flavored pale ales also encouraged the use of sugar adjuncts. After their introduction, Scottish brewers not only used sugar to brew India ales, but were more apt to use "saccharine" (sugar) in Scotch ale as well. Sugar was used as an "auxiliary, to either the copper, or the tun, the latter after the ferment has hit its peak."[4] It was added only after the peak to avoid the problem of overfoaming. (One pound of sugar per Imperial gallon, whether Fine Mauritius, Havannah or Scotch Titler, gave OG 1.036 or 1.037 or about 40 percent better extract than an equal portion of malt).[4]

Final gravities dropped along with original gravities, but both remained higher than for pale ales. Whereas English ales were (and are) generally fermented out to quarter gravity or less, Scots ales were (and many still are) only attenuated down to one-third gravity (i.e. from 1.060 [14.7 °B] down to 1.020 [5.1 °B], rather than down to 1.015 [3.8 °B]). "Scottish brewers have always tended to stop fermentation a few degrees higher," remarks Caledonian's Russell Sharp, "and use hops to mask that sweetness, so that although our ales aren't bitter, neither are they as sweet as one would expect."[15]

THE CELLARS

"When ... the heat of the gyle has been for some time stationary, and rather decreasing than otherwise, it is then necessary to remove it. And here, the practice as followed in Scotland is quite different from that adopted in England; for instead of being at once cleansed into casks, it is let down into a square of a size similar to that of the tun. Nearly the whole of the yeast ... is left in the fermenting tun. ... The gyle is then allowed to remain in the square from 12 to 36 hours, and before being run into casks, it is in fact moderately fine. When this is the case and fermentation appears quite exhausted, it is run into hogsheads, barrels, and half hogsheads. These casks are not placed upon stillions, as is the practice in England, little or no yeast being thrown up ... and require no filling up until after the lapse of a few days, when they are shived down."[4] Scots ales only underwent incidental secondary fermentation, having fermented to completion in the tun. Beer for export was flattened before being run into casks, so that excessive carbonation from incidental fermentation in the casks would not be as likely to rupture them.

Scotch ales, bottled Scottish beer, and export stout took up to a year of cellaring to come into condition,[4] largely because of the brewers' practice of beating the krauesen head back into the fermenting ale.

Beating the head drove the yeast back down into suspension so that fermentation wouldn't abate prematurely. It also reintroduced trub that had been carried up into the head back into the beer, giving to the young beer a harsh astringency. Sufficient aging would mellow these trub flavors, while the protein trub itself nurtured the remaining yeast during cellaring. Aging produced the mellow end flavors by metabolic reduction of aldehydes to alcohols, and by esterification.[25] The flavors ultimately developed by this practice are unique to wee heavy, and nicely balance the richness of the brew.[25]

Table beer, like India ale, was given no such period for mellowing. "It is not unusual to send this ale out in forty-eight hours after it has been cleansed. The Scottish brewers make no use of isinglass for finings; nor do I believe they have any occasion to employ such agents as flour and salt in order to stimulate fermentation."[4] It is no wonder that Scots table beer and twopenny was regarded with such disfavor. Reverand James Somerville termed it "thin, vapid, sour stuff under the name of sour beer" and blamed it for the working class turning to whiskey.[11]

CASKS AND TALL FONTS

Traditional Scottish serving temperatures are akin to English; the usual temperature recommended being 54 to 56 degrees F (11 to 13 degrees C). Cask conditioned ales are available in most Scottish pubs, and several are revered for their artistry in handling the casks. In the pub cellar, a cask is fitted with a soft, porous spile made of "lorit" (linden wood—called limewood in England) until carbon dioxide generation slows. A hard spile of walnut then replaces the soft one to allow a small amount of carbonation to develop in the beer. Stainless-steel Sankey-type kegs are commonly used for cask-conditioned beers. A screen shields the opening of the pick-up tube and draws the beer from an inch or so up from the bottom of the keg, to prevent sediment from fouling the beer. During conditioning, a modified tap fitted with a soft spile allows the kegs to vent.

After delivery, casks are allowed to settle for at least 24 hours upon their stillions in the pub cellars. With Sankey kegs the usual tap is attached for serving; otherwise, the hard spile is pulled so that a vacuum will not be drawn during serving by hand pump.

Very few Scottish pubs, at present or in the past, feature hand pumps. As Broughton's David Younger explains, "Scots still consider hand pumps to be English. Traditionally, we have served Scottish ales using the 'tall font.' "[15] The tall font was pioneered by Ballingall's of Dundee in the 1870s, and has been popular since at least the 1920s. Originally, a water-turbine-driven pump propelled air into the cask to push the beer up to the faucet. Currently, electric compressors are more likely to provide that pressure.

In true Scots fashion, the tall font recycles spillage directly back to the faucet. Within the font, a "sparkler" obstructs the flow so that the beer releases carbonation. Scots, unlike the English, favor a head on their ales. The flow rate of the font is adjustable, so that the barman may pour with the close, creamy head that patrons expect. Even the rare handpumps are fitted with sparklers, so that they will pour with a head on the beer. "Guinness" faucets that have become popular in America in the last decade (used in conjunction with a nitrogen and carbon dioxide mix to push the beer) have similar sparkler and flow-adjustment mechanisms to achieve this same end.

There are those who argue that a Scottish ale tastes best when served at the Bull in Cowgate, or at Diggers on Henderson Terrace, or in another traditional Scots pub, and it is true that the malty Scottish ales do no withstand the abuse of traveling with the fortitude of hoppier beers. Scottish ales such as Belhaven Export, and more especially McEwan's Export, that are exported to the United States are often badly oxidized. It is well to remember, however, that the fame of Scottish brewing more properly resides with the strong Scotch alcs, which fare somewhat better. MacAndrew's (Caledonian) Scotch Ale is the archetypal definition of maltiness when it is even reasonably fresh, and Traquair House Ale (difficult though it may be to find), is not diminished in the least by a year or more of storage.

8

Notes to Recipes

The recipes following are divided between ales characteristic of 1850, and those of the present day. You will notice that not only has the shilling greatly devalued over time, but that the final gravities of 1991 ales have dropped considerably compared with 1850 ales of similar original gravity. These are not misprints. The greatest difference between 1850 ales and 1990 ales are their final gravities; the 19th century ales end at specific gravities over twice that of the 20th century recipes. High mash temperatures are necessary to achieve these final gravities.

WATER

Any soft-to-medium-hard water will suffice for brewing Scotch ales. The worksheet below is given for those who wish to more closely approximate Edinburgh water:

- Water Treatment -

IONS	Edinburgh Water – Your Water = Treatment Column 1	– Column 2	= Column 3	1 gram of: $CaSO_4$	$MgSO_4$	$NaCl$ in 5 gallons gives:
Calcium	80 - 120 mg/L	17	83	12 mg/L		
Magnesium	10 - 25	6	9		7.4 mg/L	
Sodium	10 - 30	2	18			21 mg/L
Sulfate	70 - 140	?	105	24 mg/L	29.1 mg/L	
Carbonate	120 - 200	69 ?	60			
Chloride	30 - 60	1	44			32.1mg/L
Hardness as $CaCO_3$	225 - 350	69	150	30 mg/L	17.5 mg/L	
Total Dissolved Solids	300 - 500	115	285	41.5 mg/L	36.5 mg/L	52.9mg/L
pH	7.1 - 7.3	7.6				

Subtract column 2 from column 1, list the difference in column 3. Use the ion concentrations given in column 4 to make mineral salt adjustments.

MALT AND MASHING

Scottish maltsters still kiln malt to a color of 2.7 to 3.5 °SRM (6 to 8 °EBC), whereas English pale malts average 2.5 to 3.5 °SRM (4 to 6 °EBC). The Scots malt has a slightly fuller character. Hugh Baird & Sons kiln malt to the higher color at their Pencaitland maltings, but it is not available in America. The flavor differences between it and pale malt are slight enough that the latter can be used. Roast barley gives most of the color and flavor to today's Scotch and Scottish ales.

Malts were darker roasted than today's pale malt in older Scotch ales, and gave a very different character. The kilning methods of the early 19th century produced a dry, toasty malt that would have dominated the beer flavor. This can be imitated by using a portion of modern, darker colored "biscuit" flavored amber malt at 23 to 33 °SRM (55 °EBC), such as that malted by De Wolf-Cosyns and available through Siebel in Chicago. Any amber malt of this color will give good results. Brown malt, available from Hugh Baird & Sons/Canada malt, can be used, but at 45 to 55 °SRM (100 to 120 °EBC) it should only compose five to seven percent of the grist (i.e., less than half of the amount of amber malt that is shown in recipes). Where neither malt is available, substitute one percent roast barley with 15 to 20 percent Munich malt of 7 to 12 °SRM (15 to 25 °EBC) for the amber malt in the recipes.

Malt quantities in the recipes are based upon 68 percent mash efficiency. At 68 percent efficiency, one pound of good malt will yield one U.S. gallon of wort at SG 1.031.5. Where your experience dictates that extract efficiency will be greater or lesser than 68 percent, adjust the amount of malt accordingly. For example, for 65 percent efficiency, multiply malt quantities in the recipe by 1.05, or for 70 percent, by 0.97. As is usual, the pH of the mash should be 5.1 to 5.4 to facilitate enzyme activity and limit extraction of tannins. Kettle pH should be below 5.3, to limit extraction of coarser bittering substances.

The original gravities of the old ales, and of some of the new, are unusually high by today's standards, and those brews will require specialized treatment in the brewhouse. Mashes are very thick to protect enzyme viability. Consequently, sparge volumes are unusually high. Even so, sparging must be very efficient when brewing these strong ales. Sparging temperatures must be high to extract greater volumes of sugars from the dense mash. A temperature of 172 degrees F (78 degrees C) in the mash should be reached as quickly as possible. Use sparge liquor at 180 degrees F (82 degrees C) initially, then reduce it to 175 or 172 degrees F (79 to 78 degrees C) when the temperature in the tun itself rises to near 172 degrees F (75 degrees C).

Volumes are given for mash and sparge liquor. They take into account the amounts of water normally retained by spent grains and hops, as well as an average evaporation rate. If your experience with a recipe yields a greater or lesser volume of wort, adjust sparging and kettle-fill volumes in future brewings.

With the usual size of current day mash tuns, ales of OG 1.088 and higher will require two mashes to yield one batch of strong ale. Double mash recipes are therefore included. These also give one batch of twopenny ale, in the traditional Scottish manner. When double mashing, you need to collect the required volumes of both worts, and each at the correct density. Liquor volumes required for mashing and sparging, and wort volume and density are subjective, however. Mashing and sparging efficiency, the moisture content of the malt, the quantity of hops needed to match the quoted HBUs, and the evaporation rate achieved in any particular kettle are all variables that will affect volumes and densities. There is no possibility of quoting exact values here. Generally speaking, the wort collected from the first mash for the strong ale should be about 80 percent of the batch volume and density, that is 16 quarts at OG 1.080 to 1.082 for a five-gallon batch of 1.090 ale. The amount of liquor given for sparging the second mash can be reduced to a more usual 50 percent volume, if the balance of liquor needed to fill the kettle for the twopenny is added to the kettle separately. It should be noted that this procedure will give a better batch of twopenny.

EXTRACT MASHING

A reasonable alternative to the complexity of double-mashing and the long brew day it requires is to employ malt extract to create part of the original gravity. Recipes using malt extract syrup to raise the gravity of a manageable mash volume are given for the strongest ales. Obviously, extract may be substituted in other recipes by making a few simple calculations. Volumes of malt extract in the recipes are based upon 74 percent solids, or OG 1.034 (9 °B) per pound of syrup per gallon (SG 1.028/0.1 kilogram syrup/liter of wort). To use dry malt extract (DME) multiply all or any part of the syrup quantity by 0.756.

When brewing with malt extract, the Scottish export, heavy and special syrups from the manufacturers Glenbrew, Geordie, Ironmaster or Brewmaker should be used for those formulations; unhopped syrups should be preferred. Unhopped Munton & Fison Old Ale gives a high final gravity and a flavor appropriate for strong Scotch ale. The hopped version has too much bitterness for the Scottish style, and should be used at no greater ratio than one-to-one with unhopped Light extract, or in combination with a partial mash. If you use this particular hopped extract, reduce the AAUs called for in the recipe by 3.8 HBU per each pound of syrup added to a five-gallon batch. Other brown or mild ale extracts, or even an Oktoberfest malt extract can be used to brew these ales. In any case, only brands that are known to give a high final gravity, such as John Bull or Telford's, should be used. A good ratio would be 40 to 50 percent of a brown or amber extract, with 50 to 60 percent light extract.

As soon as some wort is run into the kettle, turn its heat on full, and lightly carmelize the wort to develop characteristic Scottish flavor and color. Although kettle carmelization gives a slightly different flavor than caramel malt does, you can substitute 5 to 10 percent crystal malt for an equal part of the pale malt.

Before the kettle comes to a boil, add about one-fourth ounce of the hops for a five-gallon batch (seven grams/20 liters), or one ounce for a barrel, to break the surface tension of the wort. This will reduce the likelihood of boiling over. The kettle boil should be 1 1/2 hours, except where excess sweet wort volume dictates a longer boil. In this case the wort should be evaporated down to 105 to 115 percent of its final volume (depending upon the evaporation rate of your kettle), and then boiled as usual for 1 1/2 hours. (Kettle evaporation is assumed to be 10 percent over a 1 1/2 hour boil; adjust kettle-fill volumes in the recipes if you know the evaporation rate in your kettle to be different.)

For ales below OG 1.044 (10.5 °B) mashing and sparging will not give the full volume of wort, and the kettle will need to be topped up before boiling. Top-up volumes are given in the recipes, so that enough water may be treated ahead of time.

HOPS

The hop perception in Scottish ales should be just enough to balance the sweetness, without intruding upon it. Hop rates, then, need to be based upon final gravity, and moderated as alcohol percentage or fermentation temperatures increase above the figures given in the recipes. Alcohol itself gives a flavor that accentuates bitterness. Fermentations at higher temperatures than those quoted will produce more of the harsh flavored fusel alcohols. Reducing the hop rate will moderate these undesirable flavors.

Bitterness Units given are for the fermentation temperatures and final gravities listed, and equal 115 percent of milligrams/liter of iso-alpha acids that can be expected to go into solution with wort of that gravity. It is practically impossible to define hopping rates beyond Bitterness Units for any beer, due to the number of variables involved. The hop rates in Homebrew Bittering Units that are given are based upon expected kettle utilizations (these are given in each recipe), but may need to be adjusted to account for your kettle. Furthermore, because Scots' ales do not require the aromatic character of fresh hops, it is possible use some of your older stock. Due to alpha-acid storage losses you may not be able to do better than to guess at their current alpha acidity. Fortunately, the bitterness of these beers is low enough that small errors will not seriously distort their character, so go ahead and use those old hops. Don't use old, spoiled hops that have a cheesy aroma.

Scots ales don't need hop aroma, and the entire hop charge should be added 45 minutes before kettle strike. If you use pellets in place of whole hops, reduce the given hop rates by 15 percent. To convert the hop quantities to pounds of alpha acid per barrel, divide the given HBUs by 1,600.

High yeast pitching rates are absolutely essential to the low-ester character of Scots ales. Homebrewers, prepare a large starter volume ahead of time, so that it is strongly fermenting at the time of pitching. Commercial breweries too, of course, will need to plan ahead in order to have an adequate volume of yeast, ready to pitch, to match the yeast rates given in the one barrel recipes.

YEAST

Use any ale yeast strain that yields low levels of esters and fusel alcohols, and gives 65 to 70 percent apparent attenuation. A Scottish strain may be preferable, but if they are not available, these ale yeasts will produce adequately similar results. A strain that produces moderate levels of diacetyl is perfectly acceptable. The Alt strain #1338 from Wyeast gives a very appropriate character, but will probably need to be beat back into the beer in order to ferment the strong ales down to their target final gravities. Wyeast's #1084 Irish Ale yeast should also give good results.

Conditioning times given are the minimum time the ale should be matured. Lengthen conditioning times if your storage temperatures are higher. On the other hand, if the ale will be filtered, conditioning times may be shortened considerably, although flavor will only improve with longer conditioning. Long conditioning is even more essential to flavor development if appreciable trub is in the ferment. Carbonate to 1 or 1.2 volumes for cask ale, to 1.5 or 1.7 volumes for keg beer, and to 1.7 or 2.0 volumes for bottled ale (volumes of CO_2 at 50 degrees F [10 degrees C]).

Points to Remember:

- Create thick mash at high temperature.
- Reach high sparge temperature quickly.
- Carmelize in the kettle.
- Hop at 45 to 60 minutes before strike.
- Pitch an adequate amount of yeast.
- Ferment at cool temperature.

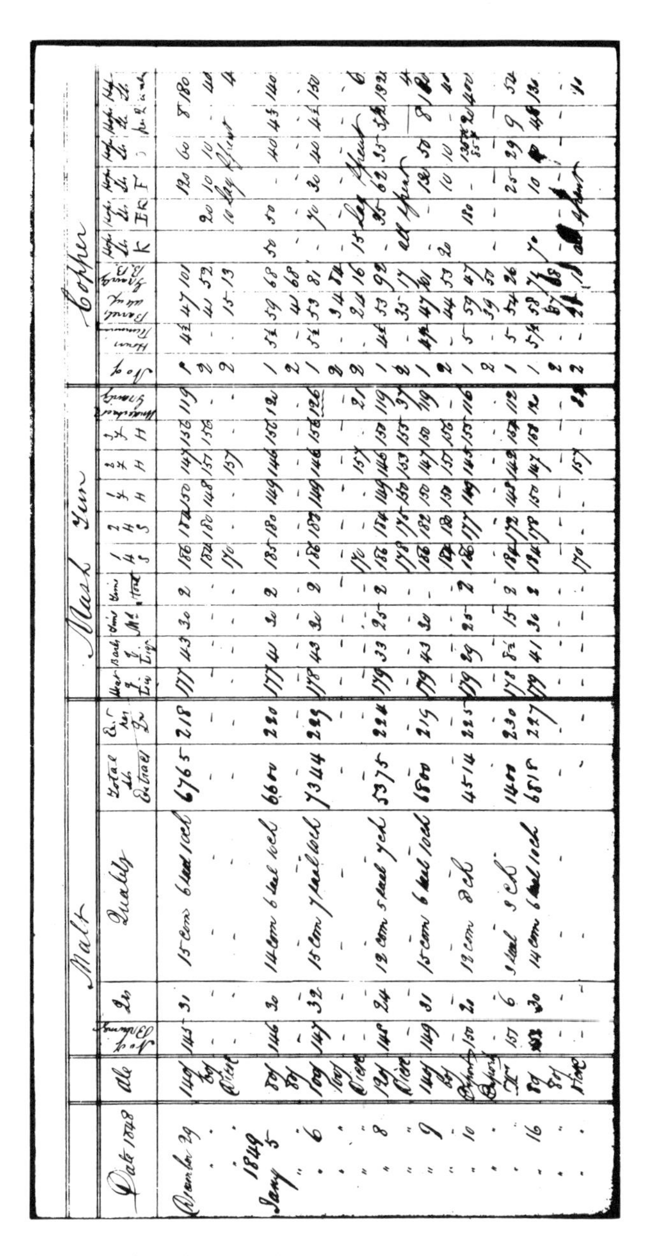

William Younger, brewing records, 1848-49. Courtesy of Scottish Brewing Archive.

1991 60 SHILLING LIGHT ALE

Single Mash

OG	1.032 (8 °B)
FG	1.010 (2.6 °B)
ABV/ABW	2.8% ABV/2.2% ABW
BU	22
HOP UTILIZATION	28%
PITCHING TEMP.	60°F (15.6°C)
TEMPERATURE LIMIT	65°F (18°C)
FERMENTATION TIME	4 days

Batch Size—	5 Gallon (Extract)	5 Gallon (Mash)	1 Barrel (Mash)
Malt Extract Syrup	4.7 lbs (2.1 kg)	—	—
Pale Malt	—	4 1/2 lbs (2 kg)	28 lbs (12.7 Carapils)
Carapils	—	8 oz (0.23 kg)	3 lbs (1.4 kg)
Roast Barley	—	1 1/4 oz (35 g)	1/2 lb (0.23 kg)
Mash Temperature	—	156°F (69°C)	156°F (69°C)
Mash Time	—	45 minutes	45 minutes
Mash Liquor	—	6 qts (5.7 L)	9 1/2 gal (0.36 hL)
Sparge Liquor	—	8 1/2 qts (8 L)	13 gal(0.49 hL)
Top Up Kettle to:	22 1/4 qts	11 qts (10.4 L)	17 1/2 gal(0.65 hL)
HBU	5.2	5.2	32.5
Yeast Volume	1 qt starter	1 qt starter	7-8 X 10^6 cells/mL
Conditioning Temp.	40-50°F (4 - 10°C)	40-50°F (4 - 10°C)	30-35°F (-1 to 2°C)
Conditioning Time	3 weeks	3 weeks	2 weeks

1991 70 SHILLING HEAVY

Single Mash

OG	1.036 (9 °B)
FG	1.012 (3.1 °B)
ABV/ABW	3.1% ABV/2.5% ABW
BU	25
HOP UTILIZATION	28%
PITCHING TEMP.	60°F (15.6°C)
TEMPERATURE LIMIT	65°F (18°C)
FERMENTATION TIME	4 days

Batch Size—	**5 Gallon (Extract)**	**5 Gallon (Mash)**	**1 Barrel (Mash)**
Malt Extract Syrup	5 lbs (2.3 kg)	—	—
Pale Malt	—	5 lbs (2.3 kg)	31 lbs (14 kg)
Carapils	—	10 oz (0.28 kg)	4 lbs (1.8 kg)
Roast Barley	—	1 1/3 oz (35 g)	1 1/3 lb (0.23 kg)
Mash Temperature	—	156°F (69°C)	156°F (69°C)
Mash Time	—	45 minutes	45 minutes
Mash Liquor	—	7 qts (6.6 L)	10 1/2 gal(0.4 hL)
Sparge Liquor	—	9 1/2 qts (9 L)	14 1/2 gal(0.55 hL)
Top Up Kettle to:	22 1/2 qts	9 1/2 qts (9 L)	15 1/2 gal(0.59 hL)
HBU	6.0	6.0	37
Yeast Volume	1 1/2 qt starter	1 1/2 qt starter	8-10 X10^6cells/mL
Conditioning Temp.	40-50°F (4 - 10°C)	40-50°F (4 - 10°C)	30-35°F (-1 to 2°C)
Conditioning Time	3 weeks	3 weeks	2 weeks

1991 80 SHILLING ALE

Single Mash

OG	1.042 (10.4 °B)
FG	1.013 (3.1 °B)
ABV/ABW	3.7% ABV/3% ABW
BU	26
HOP UTILIZATION	28%
PITCHING TEMP.	60°F (15.6°C)
TEMPERATURE LIMIT	65°F (18°C)
FERMENTATION TIME	4 - 5 days

Batch Size—	**5 Gallon (Extract)**	**5 Gallon (Mash)**	**1 Barrel (Mash)**
Malt Extract Syrup	6.1 lbs (2.8 kg)	—	—
Pale Malt	—	6.6 lbs (3.3 kg)	41 lbs (18.6 kg)
Roast Barley	—	1 1/4 oz (35 g)	1/2 lb (0.23 kg)
Mash Temperature	—	156°F (69°C)	156 °F (69°C)
Mash Time	—	45 minutes	45 minutes
Mash Liquor	—	8 qts (7.6 L)	12 1/2 gal(0.47 hL)
Sparge Liquor	—	1 qts (10 L)	17 gal(0.64 hL)
Top Up Kettle to:	22 1/2 qts	7 3/4 qts (7.3 L)	12 gal(0.45 hL)
HBU	6.2	6.2	38.4
Yeast Volume	2 qt starter	2 qt starter	10-12X10^6 cells/mL
Conditioning Temp.	40-50°F (4 - 10°C)	40-50°F (4 - 10°C)	30-35°F (-1 to 2°C)
Conditioning Time	4 weeks	4 weeks	2 weeks

1991 90 SHILLING SCOTCH ALE

Single Mash

OG	1.075 (18 °B)
FG	1.018 (4.6 °B)
ABV/ABW	7.4% ABV/5.9% ABW
BU	28
HOP UTILIZATION	25%
PITCHING TEMP.	61°F (16°C)
TEMPERATURE LIMIT	66°F (19°C)
FERMENTATION TIME	6 - 10 days

Batch Size—	5 Gallon (Extract)	5 Gallon (Mash)	1 Barrel (Mash)
Malt Extract Syrup	11 lbs (5 kg)	—	—
Pale Malt	—	12 lbs (5.4 kg)	73 1/2 lbs (33.6 kg)
Roast Barley	—	2 oz (57 g)	3/4 lb (0.34 kg)
Mash Temperature	—	150 °F (65.6°C)	150°F (65.6°C)
Mash Time	—	1 1/4 hours	1 1/4 hours
Mash Liquor	—	14 qts (13.3 L)	21 gal(0.8 hL)
Sparge Liquor	—	16 1/2 qts (15.6 L)	26 gal(1 hL)
Top Up Kettle to:	22 1/2 qts	—	—
HBU	7.5	7.5	46.4
Yeast Volume	3 qt starter	3 qt starter	16-18X 10^6 cells/mL
Conditioning Temp.	40-50°F (4 - 10°C)	40-50°F (4 - 10°C)	30-35°F (-1 to 2°C)
Conditioning Time	6 - 8 weeks	6 - 8 weeks	2 - 4 weeks

1991 120 SHILLING WEE HEAVY

Single Mash

OG	1.090 (21.5 °B)
FG	1.022 (5.5 °B)
ABV/ABW	8.8% ABV/7% ABW
BU	30
HOP UTILIZATION	22%
PITCHING TEMP.	60°F (15.6°C)
TEMPERATURE LIMIT	68°F (20°C)
FERMENTATION TIME	12 days

Batch Size—	**5 Gallon (Extract)**	**5 Gallon (Mash/Syrup)**	**1 Barrel (Mash/Syrup)**
Malt Extract Syrup	13.2 lbs (6 kg)	4.4 lbs (2 kg)	28 lbs (12.4 kg)
Pale Malt	—	7 1/2 lbs (3.4 kg)	45 1/2 lbs (20.6 kg)
Carapils	—	2 lbs (0.9 kg)	12 1/2 lbs (5.7 kg)
Roast Barley	—	1 oz (28 g)	1/2 lb (0.23 kg)
Mash Temperature	—	158°F (70°C)	158°F (70°C)
Mash Time	—	1 1/2 hours	1 1/2 hours
Mash Liquor	—	10 1/2 qts (9.9 L)	16 1/2 gal(0.62 hL)
Sparge Liquor	—	15 1/2 qts (14.7 L)	23 1/2 gal(0.9 hL)
Top Up Kettle to:	22 1/2 qts	2 qts (1.9 L)	8 1/2 gal(9.5 L)
HBU	9.1	9.1	56.6
Yeast Volume	4 qt starter	4 qt starter	18-20 X 10^6 cells/mL
Conditioning Temp.	40-50°F (4 - 10°C)	40-50°F (4 - 10°C)	30-35°C (-1 to 2°C)
Conditioning Time	2 - 3 months	2 - 3 months	4 - 6 weeks

120 SHILLING SCOTCH ALE

Double Mash

which yields two separate brews

	Strong Ale	Twopenny
OG	1.090 (21.5 °B)	1.040 (10 °B)
FG	1.022 (5.5 °B)	1.013 (3.3 °B)
ABV/ABW	8.8% ABV/7% ABW	3.5% ABV/2.8% ABW
BU	30	15
HOP UTILIZATION	22%	13% from 'spent' hops
PITCHING TEMP.	60°F (15.6°C)	60°F (15.6°C)
TEMPERATURE LIMIT	68°F (20°C)	68°F (20°C)
FERMENTATION TIME	12 days	4 - 5 days

1ST MASH:

Batch Size—	**5 Gallon (Mash)**	**1 Barrel (Mash)**
Pale Malt	8 3/4 lbs (3.98 kg)	54 lbs (24.5 kg)
Carapils	1 1/2 lbs (0.68 kg)	9 1/2 lbs (4.3 kg)
Roast Barley	1.6 oz (45 g)	0.6 lb (0.3 kg)
Mash Temperature	154°F (67.8°C)	154°F (67.8°C)
Mash Time	1 hour	1 hour
Mash Liquor	13 qts (12.3 L)	20 gal(0.76 hL)
Sparge Liquor	23 qts (21.8 L)	36 gal(1.36 hL)

Run the sweet wort into a holding vessel until 16 1/2 quarts (15.6 liters) are collected for a five-gallon batch, or 25 1/2 gallons (0.97 hectoliter) for a barrel batch. Run balance into kettle. Fill kettle to 13-quart level (12.3 liters) for five-gallon batch, or 20-gallon level (0.76 hectoliter) for barrel batch. Heat this weak wort to use for the second mash.

2ND MASH:
as above, except:

Batch Size—	**5 Gallon**	**5 Gallon (Mash)**	**1 Barrel (Mash)**
Sparge Liquor		22 qts (20.8 L)	34 gal(1.29 hL)

From the second mash, run 6 quarts (5.7 liters) to the 120 shilling ale for a five-gallon batch. For a barrel batch, run 9 1/2 gallons to the 120 shilling. The object is to collect 22 1/2 quarts (21.3 liters) volume for a five-gallon batch, or 35 gallons (1.33 hectoliters) volume for a barrel batch. The remaining sweet wort from sparging should be run off to a second holding vessel for the twopenny. The wort volume collected for the twopenny should be 22 quarts (18 liters) for a five-gallon batch, or 34.1 gallons (1.29 hectoliters) for a barrel batch. Transfer the 120 shilling wort to the kettle and boil it for 1 1/2 hours, adding 10 percent of the hops at the beginning of the boil, and all the rest at 45 minutes.

STRONG ALE

Batch Size—	**5 Gallon Mash**	**1 Barrel Mash**
HBU	9.1	56.6
Yeast Volume	4 qt starter	18-20 X10^6 cells/mL
Conditioning Temp.	40-50°F (4 - 10°C)	30-35°C (-1 to 2°C)
Conditioning Time	2 - 3 months	4 - 6 weeks

After the 120 shilling ale wort has been run out of the kettle, refill it with the twopenny wort, on top of the spent hops and begin boiling the twopenny wort.

TWOPENNY

Batch Size—	**5 Gallon Mash**	**1 Barrel Mash**
Yeast Volume	1 1/2 qt starter	8-10 X 10^6 cells/mL
Conditioning Temp.	40-50°F (4 - 10°C)	30-35°C (-1 to 2°C)
Conditioning Time	4 weeks	2 weeks

1991 140 SHILLING WEE HEAVY

Single Mash

OG	1.100 (23.7 °B)
FG	1.028 (7.1 °B)
ABV/ABW	9.3% ABV/7.4% ABW
BU	40
HOP UTILIZATION	20%
PITCHING TEMP.	60°F (15.6°C)
TEMPERATURE LIMIT	68°F (20°C)
FERMENTATION TIME	14 days

Batch Size—	5 Gallon (Extract)	5 Gallon (Mash/Syrup)	1 Barrel (Mash/Syrup)
Malt Extract Syrup	14.6 lbs (6.6 kg)	6.6 lbs (3 kg)	41 lbs (18.6 kg)
Pale Malt	—	6 1/4 lbs (2.8 kg)	8 1/2 lbs (17.5 kg)
Carapils	—	2 1/2 lbs (1.1 kg)	15 1/2 lbs (7 kg)
Roast Barley	—	3/4 oz (21 g)	0.3 lb (0.14 kg)
Mash Temperature	—	158°F (70°C)	158°F (70°C)
Mash Time	—	1 1/4 hours	1 1/4 hours
Mash Liquor	—	9 qts (8.5 L)	14 gal(0.53 hL)
Sparge Liquor	—	14 qts (13.3 L)	22 gal(0.83 hL)
Top Up Kettle to:	23 1/2 qts	5 1/4 qts (5 L)	7 3/4 gal(0.29 hL)
HBU	13.3	13.3	82.8
Yeast Volume	4 qt starter	4 qt starter	18-20 X 10^6 cells/mL
Conditioning Temp.	40-50°F (4 - 10°C)	40-50°F (4 - 10°C)	30-35°F (-1 to 2°C)
Conditioning Time	2 - 3 months	2 - 3 months	4 - 6 weeks

140 SHILLING WEE HEAVY

Double Mash

which yields two separate brews

	Strong Ale	Twopenny
OG	1.100 (23.7 °B)	1.040 (10 °B)
FG	1.028 (7.1 °B)	1.013 (3.3 °B)
ABV/ABW	9.3% ABV/7.4% ABW	3.5% ABV/2.8% ABW
BU	40	30
HOP UTILIZATION	20%	15% from 'spent' hops
PITCHING TEMP.	60°F (15.6°C)	60°F (15.6°C)
TEMPERATURE LIMIT	68°F (20°C)	68°F (20°C)
FERMENTATION TIME	14 days	4 - 5 days

1ST MASH:

Batch Size—	5 Gallon Mash	1 Barrel Mash
Pale Malt	9 lbs (4.1 kg)	56 lbs (25.4 kg)
Carapils	2 lbs (0.98 kg)	12 1/2 lbs (5.7 kg)
Roast Barley	1 1/2 oz (43 g)	1/2 lb (0.23 kg)
Mash Temperature	155°F (68.3°C)	155°F (68.3°C)
Mash Time	1 hour	1 hour
Mash Liquor	13 1/2 qts (12.8 L)	21 gal(0.8 hL)
Sparge Liquor	23 1/4 qts (22 L)	36 gal(1.36 hL)

Run the sweet wort into a holding vessel until 16 quarts (15.1 liters) are collected for a five-gallon batch, or 25 gallons (0.95 hectoliter) for a barrel batch. Run balance into kettle. Fill kettle to 13 1/2 quart level (12.8 liters) for five-gallon batch, or 21 gallon level (0.8 hectoliter) for barrel batch. Heat this weak wort to use for the second mash.

2ND MASH
as above, except:

Batch Size—	5 Gallon Mash	1 Barrel Mash
Sparge Liquor	23 1/4 qts (22 L)	35 gal(1.33 hL)

After second mash, run 6 3/4 quarts (6.4 liters) to the 140 shilling ale for a five-gallon batch. For a barrel batch, run 10 gallons (0.39 hectoliter) to the 140 shilling. The object is to collect 22 3/4 quarts (21.5 liters) volume for a five-gallon batch, or 35 gallons (1.33 hectoliters) volume for a barrel batch. The remaining sweet wort from sparging should be run off to a second holding vessel for the twopenny. The wort volume collected for the twopenny should be 22 quarts (18 liters) for a five-gallon batch, or 34.1 gallons (1.29 hectoliters) for a barrel batch. Transfer the 140 shilling wort to the kettle and boil it for 1 1/2 hours, adding 10 percent of the hops at the beginning of the boil, and all the rest at 45 minutes.

Batch Size—	5 Gallon Mash	1 Barrel Mash
HBU	13.3	82.8
Yeast Volume	4 qt starter	18-20X10^6 cells/mL
Conditioning Temp.	40-50°F (4 - 10°C)	30-35°C (-1 to 2°C)
Conditioning Time	2 - 3 months	4 - 6 weeks

After the 140 shilling ale wort has been run out of the kettle, refill the copper with the twopenny wort, on top of the spent hops and commence boiling the twopenny wort.

TWOPENNY

Batch Size—	5 Gallon Mash	1 Barrel Mash
Yeast Volume	1 1/2 qt starter	8-10X10^6 cells/mL
Conditioning Temp.	40-50°F (4 - 10°C)	30-35°C (-1 to 2°C)
Conditioning Time	4 weeks	2 weeks

1850 60 SHILLING SCOTTISH ALE

Single Mash

OG	1.074 (17.9 °B)
FG	1.030 (7.5 °B)
ABV/ABW	5.7% ABV/4.6% ABW
BU	28
HOP UTILIZATION	25%
PITCHING TEMP.	60°F (15.6°C)
TEMPERATURE LIMIT	65°F (18°C)
FERMENTATION TIME	6 - 10 days

Batch Size—	5 Gallon (Extract)	5 Gallon (Mash)	1 Barrel (Mash)
Malt Extract Syrup	10.9 lbs (4.94 kg)	—	—
Pale Malt	—	10.3 lbs (4.67 kg)	64 lbs (29 kg)
Amber Malt	—	1 1/2 lbs (0.68 kg)	9 1/4 lbs (4.2 kg)
Mash Temperature	—	156°F (69°C)	156°F (69°C)
Mash Time	—	45 minutes	45 minutes
Mash Liquor	—	11 1/2 qts (10.9 L)	17 3/4 gal(0.67 hL)
Sparge Liquor	—	18 1/2 qts (17.5 L)	29 1/4 gal(1.11 hL)
HBU	7.5	7.5	46.4
Yeast Volume	3 qt starter	3 qt starter	16-18 X 10^6cells/mL
Conditioning Temp.	40-50°F (4 - 10°C)	40-50°F (4 - 10°C)	30-35°F (-1 to 2°C)
Conditioning Time	8 weeks	8 weeks	4 weeks

1850 80 SHILLING EXPORT

Single Mash

OG	1.088 (2.1 °B)
FG	1.033 (13.5 °B)
ABV/ABW	7.1% ABV/5.7% ABW
BU	40
HOP UTILIZATION	22%
PITCHING TEMP.	55°F (12.8°C)
TEMPERATURE LIMIT	65°F (18°C)
FERMENTATION TIME	14 days

Batch Size—	**5 Gallon (Extract)**	**5 Gallon (Mash/Syrup)**	**1 Barrel (Mash/Syrup)**
Malt Extract Syrup	12.9 lbs (5.9 kg)	4.4 lbs (2 kg)	28 lbs (12.4 kg)
Pale Malt	—	8 lbs (3.6 kg)	49 lbs (22.2 kg)
Amber Malt	—	1 1/4 lb (0.57 kg)	7 1/2 lbs (3.4 kg)
Mash Temperature	—	156°F (69°C)	156°F (69°C)
Mash Time	—	45 minutes	45 minutes
Mash Liquor	—	8 3/4 qts (8.3 L)	9 1/4 gal(0.35 hL)
Sparge Liquor	—	15 qts (14.2 L)	23 gal(0.87 hL)
Top Up Kettle to:	23 1/4 qts	3 1/2 qts (3.3 L)	9 3/4 gal(0.37 hL)
HBU	12.1	2.1	75.2
Yeast Volume	4 qt starter	4 qt starter	18-20X10^6 cells/mL
Conditioning Temp.	40-50°F (4 - 10°C)	40-50°F (4 - 10°C)	30-35°F (-1 to 2°C)
Conditioning Time	3 months	3 months	4 - 6 weeks

80 SHILLING EXPORT ALE

Double Mash

which yields two brews

	Strong Ale	Twopenny
OG	1.088 (21 °B)	1.040 (10 °B)
FG	1.033 (13.5 °B)	1.013 (3.3 °B)
ABV/ABW	7.1% ABV/5.7% ABW	3.5% ABV/2.8% ABW
BU	40	29
HOP UTILIZATION	22%	13% from 'spent' hops
PITCHING TEMP.	55°F (12.8°C)	60°F (15.6°C)
TEMPERATURE LIMIT	65°F (18°C)	68°F (20°C)
FERMENTATION TIME	15 days	4 - 5 days

1ST MASH:

Batch Size—	5 Gallon Mash	1 Barrel Mash
Pale Malt	9 lbs (4.1 kg)	55 1/2 lbs (25.2 kg)
Amber Malt	1 1/4 lbs (0.57 kg)	7 3/4 lbs (3.5 kg)
Mash Temperature	156°F (69°C)	156°F (69°C)
Mash Time	45 minutes	45 minutes
Mash Liquor	10 qts (9.5 L)	15 1/4 gal(0.58 hL)
Sparge Liquor	23 1/4 qts (22 L)	36 gal(1.36 hL)

Run the sweet wort to a holding vessel until 16 1/2 quarts (15.6 liters) are collected for a five-gallon batch, or 25 1/2 gallons (0.97 hectoliter) for a barrel batch. Run balance of runoff to kettle. Fill kettle to 10-quart level (9.5 liters) for five-gallon batch, or 15 gallons (0.58 hectoliter) for each barrel batch. Heat this weak wort to use for the second mash.

2ND MASH
as above except:

Batch Size—	**5 Gallon Mash**	**1 Barrel Mash**
Sparge Liquor	25 qts (23.7 L)	39 gal(1.48 hL)

From the second mash, run 6 1/4 quarts (5.9 liters) to the 80 shilling ale for a five-gallon batch. For a barrel batch, run 9 3/4 gallons (0.37 hectoliter) to the 80 shilling. The object is to collect 22 3/4 quarts (21.5 liters) volume for a five-gallon batch, or 35 1/4 gallons (1.33 hectoliters) volume for a barrel batch. The remaining sweet wort from sparging should be run off to a second holding vessel for the twopenny. The wort volume collected for the twopenny should be 22 quarts (18 liters) for a five-gallon batch, or 34.1 gallons (1.29 hectoliters) for a barrel batch. Transfer the 80 shilling wort to the kettle and boil it for 1 1/2 hours, adding 10 percent of the hops at the beginning of the boil, and all the rest at 45 minutes.

STRONG ALE

Batch Size—	**5 Gallon Mash**	**1 Barrel Mash**
HBU	12.1	75.2
Yeast Volume	4 qt starter	18-20X10^6 cells/mL
Conditioning Temp.	40-50°F (4 - 10°C)	30-35°C (-1 to 2°C)
Conditioning Time	3 months	4 - 6 weeks

After the 80 shilling ale wort has been run out of the kettle, refill the copper with the twopenny wort, on top of the spent hops and commence boiling the twopenny wort.

TWOPENNY

Batch Size—	**5 Gallon Mash**	**1 Barrel Mash**
Yeast Volume	1 1/2 qt starter	8-10X10^6 cells/mL
Conditioning Temp.	40-50°F (4 - 10°C)	30-35°C (-1 to 2°C)
Conditioning Time	4 weeks	2 weeks

1850 140 SHILLING EDINBURGH ALE

Single Mash

OG	1.125 (29 °B)
FG	1.055 (13.5 °B)
ABV/ABW	9% ABV/13.5% ABW
BU	60
HOP UTILIZATION	20%
PITCHING TEMP.	50°F (10°C)
TEMPERATURE LIMIT	62°F (17°C)
FERMENTATION TIME	21 days

Batch Size—	**5 Gallon (Extract)**	**5 Gallon (Mash/Syrup)**	**1 Barrel (Mash/Syrup)**
Malt Extract Syrup	18.4 lbs (8.4 kg)	8.4 lbs (3.8 kg)	52 lbs (23.6 kg)
Pale Malt	—	9 1/2 lbs (4.3 kg)	58 1/2 lbs (26.5 kg)
Amber Malt	—	1 1/2 lbs (0.68 kg)	9 lbs (4.1 kg)
Mash Temperature	—	158°F (70°C)	158°F (70°C)
Mash Time	—	2 hours	2 hours
Mash Liquor	—	10 1/2 qts (9.9 L)	16 1/4 gal(0.62 hL)
Sparge Liquor	—	18 qts (17 L)	27 3/4 gal(1 hL)
Top Up Kettle to:	24 qts	2 qts (1.9 L)	3 gal(0.11 hL)
HBU	20	20	124
Yeast Volume	4 qt starter	4 qt starter	20-22X10^6 cells/mL
Conditioning Temp.	40-50°F (4 - 10°C)	40-50°F (4 - 10°C)	30-35°F (-1 to 2°C)
Conditioning Time	4 - 6 months	4 - 6 months	2 - 3 months

140 SHILLING EDINBURGH ALE

Double Mash

which yields two separate brews

	Strong Ale	Twopenny
OG	1.125 (29 °B)	1.040 (10 °B)
FG	1.055 (13.5 °B)	1.013 (3.3 °B)
ABV/ABW	9% ABV/7.2% ABW	3.5% ABV/2.8% ABW
BU	60	45
HOP UTILIZATION	20%	15%, from 'spent' hops
PITCHING TEMP.	50°F (10°C)	60°F (15.6°C)
TEMPERATURE LIMIT	62°F (17°C)	68°F (20°C)
FERMENTATION TIME	21 days	4 - 5 days

1ST MASH:

Batch Size—	5 Gallon Mash	1 Barrel Mash
Pale Malt	11 3/4 lbs (5.3 kg)	72 1/2 lbs (32.9 kg)
Amber Malt	1 1/2 lbs (0.68 kg)	9 1/4 lbs (4.2 kg)
Mash Temperature	158°F (70°C)	158°F (70°C)
Mash Time	2 hours	2 hours
Mash Liquor	13 qts (12.3 L)	20 gal(0.76 hL)
Sparge Liquor	24 qts (22.7 L)	37 gal(1.4 hL)

Run the sweet wort to a holding vessel until 15 1/4 quarts (14.4 liters) are collected for a five-gallon batch, or 23 3/4 gallons (0.9 hectoliters) for a barrel batch. Run balance into kettle. Fill kettle to 13-quart level (12.3 liters) for five-gallon batch, or 20 1/2-gallon level (0.76 hectoliter) for a barrel batch. Heat this weak wort to use for the second mash.

2ND MASH
as above, except:

Batch Size—	5 Gallon Mash	1 Barrel Mash
Sparge Liquor	26 qts (24.6 L)	40 gal(1.5 hL)

From the second mash, run 8 quarts (7.6 liters) to the 140 shilling ale for a five-gallon batch. For a barrel batch, run 12 gallons (0.46 hectoliter) to the 140 shilling. The object is to collect 23 1/4 quarts (22 liters) volume for a five-gallon batch, or 36 gallons (1.36 hectoliters) volume for a barrel batch. The remaining sweet wort from sparging should be run off to a second holding vessel for the twopenny. The wort volume collected for the twopenny should be 22 quarts (18 liters) for a five-gallon batch, or 34.1 gallons (1.29 hectoliters) for a barrel batch. Transfer the 140 shilling wort to the kettle and boil it for 1 1/2 hours, adding 10 percent of the hops at the beginning of the boil, and all the rest at 45 minutes.

Batch Size—	5 Gallon Mash	1 Barrel Mash
HBU	20	124
Yeast Volume	4 qt starter	20-22X10^6 cells/mL
Conditioning Temp.	40-50°F (4 - 10°C)	30-35°C (-1 to 2°C)
Conditioning Time	4 - 6 months	2 - 3 months

After the 140 shilling ale wort has been run out of the kettle, refill it with the twopenny wort, on top of the spent hops and commence boiling the twopenny wort.

TWOPENNY

Batch Size—	5 Gallon Mash	1 Barrel Mash
Yeast Volume	1 1/2 qt starter	8-10X10^6 cells/mL
Conditioning Temp.	40-50°F (4 - 10°C)	30-35°C (-1 to 2°C)
Conditioning Time	4 weeks	2 weeks

Appendix A: Breweries

Alloa Brewery Co. Ltd. The Brewery, Whins Road, Alloa, Clackmannanshire, Central FK10 3RB, Scotland, (0259) 723539. Office: Orchard Brae House, 30 Queensbury Road, Edinburgh, Scotland. Established in 1810. Archibal Arrol purchased the brewery in 1866. Alloa was taken over by Inde Coope, which was bought out by Vaux/Lorimer; they, in turn, were swallowed up by Allied Brewers in 1980. The modern brewery on the outskirts of Alloa is the third largest in Scotland, and stands in sharp contrast to the town's other remaining brewery, Maclay's Thistle Brewery.

Products

Arrol's 80 Shilling. OG 1.042, 4.2% ABV. Amber-brown in color and translucent. Has a light flavor that is slightly fruity and bitter, with a deep, buttery accent to its mild sweetness. Medium finish.

Archibal Arrols 70 Shilling. Also known as Keg Heavy. OG 1.037. Hoppier. Probably the same beer as Dryborough Heavy.

Dryborough Heavy 70 Shilling. OG 1.034 to 38. Deep copper-brown color. Slightly fruity, very buttery aroma. Dense head with good cling. Light roastiness, sweetness and diacetyl with mild bitterness. Low gas, medium-light body. Short finish with faint lingering dryness. Dryborough's Craigmillar Brewery in Edinburgh was closed in 1987 when Watney-Mann was taken over by Allied, but the brand name survives and is widely available.

Alloa also markets Inde Coope Burton Ale (OG 1.048, 4% ABV). It is a real ale.

Belhaven Brewery Co. Ltd. Dunbar, Lothian Eh42 IRS, Scotland, (0368) 62734. Owns 40 pubs and hotels. The brewery employs three brewers, and a total of 24 people. The brewery produces 600 barrels (of 36 Imperial gallons each) per week; real ale production is 100 barrels per week.

The brewery is built around a 13th century garden; some of the vaulted cellars date back to the 15th century, when it was still a monastery brewery. Belhaven remains Scotland's oldest independent brewery. In 1719 the John Johnstone family took over the premises for use as a "Publick Brewery." In 1815 Ellis Dudgeon took over; the brewery continued to trade under his name until 1973. The brewery was late in being modernized; no refrigeration and carbonation equipment was used until 1948. Since that time it has been thoroughly modernized, including a 120-barrel stainless-steel calandria kettle, conical fermenters and computer controlled fermentation. Cask conditioned ales, however, are still fermented out in four days, starting at 58 degrees F (11 degrees C), in open fermenters. The ales are racked into casks with finings on the eighth day. They also bottle for Broughton, Alloa, Tennent's and Traquair House.

Kent Goldings are the hops used almost exclusively in Belhaven's ales: 6 1/2 to 7 percent glucose (corn sugar) is added to the kettle.

Belhaven Brewery. Photo by Greg Noonan.

Products

Belhaven 90 Shilling Strong Ale. OG 1.070, FG 1.014, 7.25% ABV. Also bottled as Fowler's Wee Heavy and Maclays Strong Ale. Malty with roasty overtones. Alcohols dominate the flavor without being overbearing. A 60 barrel batch is brewed four times a year. Available in 180 milliliter bottles, or on tap at half a dozen pubs.

80 Shilling Export. OG 1.041, FG 1.010, 3.9% ABV. Brewed with 5 percent crystal malt and 2 percent roast barley. Malty, toffee, butterscotch and raisinlike flavors are nicely balanced by subtle hopping. A soft, full-bodied, heavy ale. Soft crystal finish gives way to a short and only-lightly-bitter finish. Exported to (and well loved in) the United States as Belhaven Scottish Ale.

70 Shilling Heavy. OG 1.035, FG 1.007, 3.3% ABV, 28 IBU. Gassy, light-flavored and hoppy, with a smooth roasty/caramel character. Even the cask version is filtered and pasteurized. Recently being kegged with less carbonation (1.2 volumes instead of Belhaven's standard 1.7)

60 Shilling Light. OG 1.031, FG 1.010, 2.8% ABV. Caramel color added at kegging. Filtered version marketed as Keg Light. Increasingly difficult to find. Dark, malty and mild.

Special. OG 1.037. Special pale ale version of the 70 shilling.

Belhaven occasionally brews an export version at OG 1.056 (4.2% ABV) for Texas. The brewery also produces Bass Light (OG 1.031), and the Tennents beers, Lincoln Green, Organic and Rheinheitsgebot Lager as well as their own house lagers, the Premium (OG 1.035 to 1.040) and the Export (OG 1.048). Both are fermented down to between FG 1.004 and 1.005.

The Bobbin Inn. Bridge of Don, Aberdeen, Scotland. Alloa's second brewpub. Its 1992 opening was announced during the celebration of Auld Reekie's 500th brew at the Rose Street Brewery.

Borve Brewhouse. Ruthven, Aberdeenshire, AB5 4SR, Scotland, (046 687) 343. Owned by James Hughes. Moved in 1989 from Isle of Lewis (Est. 1983) to beside the A96, Huntley-Keith Road, in a converted schoolhouse.

Products

Extra Strong. OG 1.085, 10% ABV. Ruby red, oak-matured strong ale.

Bishop Elphinstone's Ale. OG 1.053, 5% ABV.

Borve Heavy Ale. OG 1.040, 3.7% ABV.

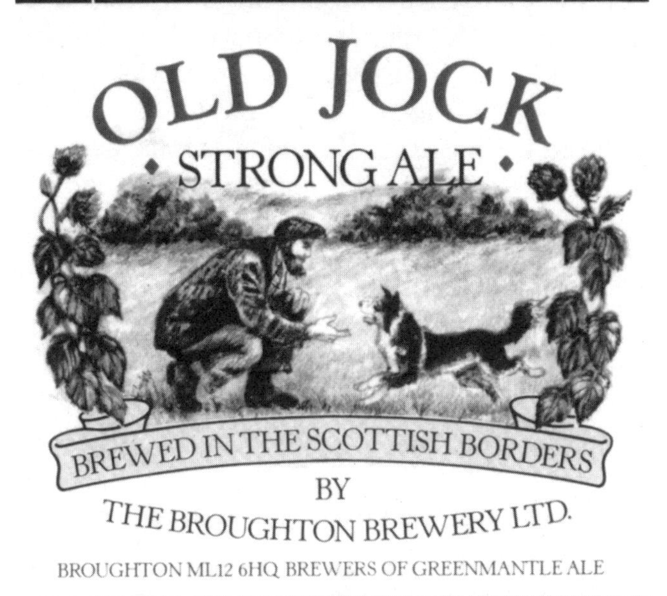

Label courtesy of The Broughton Brewery.

Broughton Brewery Ltd. Broughton, Peeblesshire, Borders ML12 6HQ, Scotland, (089 94) 345. Established May 14th, 1980. David Younger, formerly of Scottish and Newcastle, has been the Managing Director and force behind this successful 60-barrel microbrewery. Mr. Younger is a seventh-generation brewer, and the fourth to start a brewery. He is descended from George Younger of Alloa and Robert Younger of the St. Ann's Brewery in Edinburgh. The brewery has the smallest tied-house estate in Scotland, consisting of only the New Bizaar in Dumfries.

Brougton's beers, like the "Special" Pale ales of other Scottish breweries, are a hybrid between English bitter and traditional Scottish ale; David Younger considers them to be "border beers." Pale malt, roast barley and flaked maize make up the usual grist composition. Roast barley is less than 1 percent of the grist in all their beers except Old Jock. Kettle carmelization is negligible. Gypsum is added to the soft well water; Old Jock has the greatest proportion of added gypsum. Fuggles and Goldings pellets constitute the bulk of the hops, with smaller amounts of Northdown.

Products

Old Jock Strong Ale. OG 1.066 to 72, FG 1.016, 6.7% ABV. Named for Old Jock, in the story *Greyfriars' Bobby.* Deep, red-brown color. Mild aroma. Full, rounded syrupy body. Rich roastiness balances ethanol and propanol in the aroma. Very lightly hopped. Deep maltiness dominates a faintly bitter background, and the finish is all of light alcohol. Only occasionally found on draft, the bottled version is well distributed.

Greenmantle 80 Shilling Export. Formerly Merlin's, ESB. OG 1.042 to 1.046, FG 1.009.5, 4.25% ABV. Light copper color. Light fruitiness comes to the fore in the aroma. Its light, buttery, and bitter flavor is nicely balanced. Highly carbonated. Cleanly bitter dry finish.

Greenmantle Ale. OG 1.036 to 1.040, FG 1.009, 4% ABV. Named for closing paragraph in the John Buchan book of the same name, "Then I knew that the prophecy had been true, and that their prophet had not failed them. The long looked-for revelation had come. Greenmantle had appeared at last to an awaiting people." Transparent, brown-tinted bronze color. Mild, sweet, slightly hoppy aroma. Grapes and cherries stand out from a complex sweetness that is syrupy and caramellike. Noticeable diacetyl. Very syrupy-and-full mouthfeel. Short finish with very subtle dry bitterness. Meant to be "a balance between hoppy English ales and the heavier maltiness of Scottish beers."

Broughton Special Bitter. The same beer, but dry hopped.

Broughton 60 Shilling Ale. OG 1.036, FG 1.008, 4% ABV. Keg only. "Hot butter" predominates in the aroma. Overly fruity, slightly vinegary, with a short finish.

Calendonian Brewing Co. Ltd. Slateford Road, Edinburgh, Lothian, EH11 1PH, Scotland, (031) 337 1286. The 19th century brewery passed from Lorimer & Clark into conglomeration and back to independence again.

There were 30 breweries operating in Edinburgh in 1869 when George Lorimer and Robert Clark built their Caledonian brewery in the "charmed circle" on Slateford Road (so called because of the quality of the well water). The brewery loaded 1,000 hogsheads of ale per week onto railcars at their siding on the Caledonian Railway's main line; Northeast England was the largest market for their Lorimer's Scotch Ale. Merman India Pale Ale, Merman Imperial Stout, Brown Stout and Double Brown Stout were also brewed and bottled. Vaux Breweries of Sunderland acquired the brewery, with its maltings, coopers' shop and 70 employees, in 1919. The maltings and the coopers' shop remained until 1964; Vaux, while still producing 1,600 barrels a week, mostly for consumption out of Scotland, announced that they would close the brewery in May of 1986. Director Russell Clark led a 1987 management buyback that has preserved this working museum. The brewery has 18 employees and serves 50 free-trade pubs. Its brewhouse boasts the last open, direct-fired coppers in the British Isles, one of which dates back to 1869.

Brewing liquor is moderately hardened to facilitate mashing and fermentation, and Golden Promise malted barley composes most of the grist. Surprisingly, their very typical Scottish ales are not brewed with roast barley. The 70 shilling uses chocolate, crystal and amber malts, and torrified wheat for flavoring. The percentage of wheat increases and the colored malts decrease in the stronger ales. The brewery does not use sugar.

Fuggles and Goldings bitter and flavor the ales; the ales ferment in open copper and stainless-steel fermenters for seven days (12 to 14 days for the Edinburgh Strong Ale) at temperatures in the 60s F. The beer is skimmed after 30 hours, at an original gravity of 12 (3 °B) above final gravity. Fermentation is stopped by dropping the temperature in order to precipitate the yeast at a specific gravity 0.5 (0.1 °B) above final gravity. The yeast strain itself has never been replaced, having been successively recropped since Lorimer and Clark established the brewery. The ales are racked to tanks or casks in the cellars for seven days or more. The casks of ale condition naturally in the cool cellars before being fined. Even Caledonian's kegged ales aren't artificially carbonated, or filtered. Beer for bottling is trucked to Sam Smith's in Yorkshire.

Products

Edinburgh Strong Ale. OG 1.080, FG 1.016 to 18, 70 to 72 IBU, 8% ABV. Roast barley is used in this wee heavy, and torrified wheat helps create its depth of flavor. Bronze colored, with a tan head that laces the glass. The full aroma is of hops, roast barley and malt. The flavor begins very sweet, increases in spiciness and ends with a long dry and warming finish. Full-bodied and its richness masks its strength. Labeled as McAndrew's Scotch Ale in the United States.

Merman XXX. OG 1.052, FG 1.014, 4.1% ABV. A heavy, dark brown ale available only in casks. Merman is a brewery trademark from 1890.

Caledonian 80 Shilling Export. OG 1.043, FG 1.011 to 12, 4.1% ABV, 38 IBU. Wheat composes 6 percent of the grist, and crystal and amber malts up to another 6 percent. Faint, burnished copper-to-brown hue. Rich, creamy head.

Dry-hop dominates the aroma. Its flavor is a masterful balance of hoppiness, maltiness, bitterness, sweetness, caramel, dry-hop, butterscotch and roast barley. The 80 shilling is extremely chewy without being full-bodied and finishes with a lingering hop bitterness that is perfectly complementary.

Caledonian 70 Shilling. OG 1.034 to 36, FG 1.009.5 to 1.010, 28 to 30 IBU, 3.5% ABV. Sold in England as Lorimer's Best Scotch Ale. Caledonian has increased the hopping from 22 IBUs since Lorimer & Clark last brewed this ale, but its high final gravity and maltiness moderate the bitterness. Brown color that sparkles with black and amber. Late hopping gives it a pleasant hoppy aroma. The flavor is well-balanced, soft and malty.

Caledonian 60 Shilling Ale. OG 1.032, FG 1.008, 3.2% ABV. Available in cask only.

R. & D. Deuchars India Pale Ale. 34 to 36 IBU, 3.8% ABV. Unfiltered. Brewed with less colored malts than the Scottish ales. It is a Scottish version of an IPA, however, and the hop character is appropriately subdued. The ale is late-hopped rather than dry-hopped. It has a very clean, alcohol-predominant aroma. The flavor is beautifully balanced, and a light body complements its clean character.

Golden Promise Organic Beer. OG 1.050, FG 1.012 to 14, 5% ABV. Naturally brewed with organically grown barley and hops, the latter a very floral Hallertauer-type from New Zealand. A very pleasant, malty ale.

Porter. OG 1.036 has also been brewed.

Harviestoun Brewery Ltd. Dollarfield Farm, Dollar, Clackmannanshire, Central, FK14 7LX, Scotland, (025 94) 2141. Ken Booker and E. Morris. Established 1985. Seven barrel brewery in a former dairy set below the Ochil hills on the River Devon near Stirling.

Products

New Year Ale. OG 1.090, 9.2% ABV in 1990. 1.091 in 1991. Their winter brew.

Old Manor. OG 1.050, 5.1% ABV. A dark ale with roast malt flavor predominating.

Harviestoun 80 Shilling. OG 1.041, 4.1% ABV. Straw-colored, full-bodied ale.

Waverly Ale 70 Shilling. OG 1.036, 3.6% ABV. Brewed for the Malt & Shovel pubs of Edinburgh, may no longer be available.

Maclay & Co. Ltd. Thistle Brewery, Alloa, Clackmannanshire, Central, FK10 1ED, Scotland, (0259) 723387. Scotland's other remaining independent brewery, founded by James Maclay at the old Mills Brewery in 1830. Real ale is served in 14 of its 25 tied houses.

The present Thistle Brewery was constructed in 1870 over pre-existing cellars. The brewery today seems little updated since then; the original wood-cabinet roller mill, Steele's masher, hemispherically-domed copper tuns, 1871 kettles and copper-lined wooden, open circular fermenters are still in daily use. Its picturesque jumble of Victorian buildings fill most of one side of Old High Street. The brewery employs its very minimal advertising judiciously. The fact that it can't afford to market itself may be good news; Maclays probably survived the century of takeovers because of its poor financial performance and its paltry tied-house estate (25 at present, many of them very modest locals).

Managing Director George King is expected to preserve Maclays heritage, and Head Brewer Duncan Kellock (11 years) and Production Manager Charlie Ritchie (42 years, former Head Brewer) brew much as Alexander Fraser did when he took over the brewery in 1896. Fraser was considered a very scientific brewer in his day. Recipes have not changed substantially since, being handed down from brewer to brewer. The brewery patented Oat Malt Stout in 1914, when war made malt a scarce commodity. When the maltings were retired in 1958, the stout was dropped, since the malted oats aren't available elsewhere.

Water from the brewery's well, is treated with gypsum, calcium chloride and magnesium sulfate. Golden Promise barley malt is purchased from Bass/Alloa and from Simpson's; roast barley gives the color and flavor to the brews. Their stouts use caramel, linseed, licorice and salt in addition. Fuggles are used for most of the hops, with 5 percent or so Canadian Bramling Cross for aroma.

From the brewhouse, all the wort is collected at 61.5 degrees F (16.5 degrees C) in a circular fermenter by 3 p.m.

Maclays' stone buildings of 1870. Photo by Greg Noonan.

The temperature rises up to 66 degrees F (19 degrees C), and sometimes to 68 degrees F (20 degrees C). The first skimming is 41 hours after pitching. After three subsequent skimmings, the temperature is dropped to 56 degrees F (13.5 degrees C) over two days. The head is always skimmed away at 8 a.m. on the third day from pitching. When fermentations vary, temperatures are altered and the end-point varies but the skimming schedule is followed religiously. Beers are filtered through diatomaceous earth at racking to casks or conditioning tanks on the sixth day. They are fined and held for 10 days at 32 degrees F (0 degrees C). Keg beer is centrifuged, then goes through rough and then sterile sheet filtering. After carbonation to between 1.3 and 1.8 volumes, and then seven days' conditioning, the beer is flash pasteurized.

Maclays stopped bottling in 1983; Belhaven now bottles for them.

Products

Scotch Ale. OG 1.050, FG 1.014, 5% ABV. Newer, golden-colored pale ale, with bitter edge and rich sweetness. Two-thirds of production is cask conditioned; the rest is kegged.

Maclays Export 80 Shilling. OG 1.040, FG 1.013, 4% ABV. As the Maclays slogan has it, "Real beer drinkers will recognize it in the dark." Pale, burnished-copper color. Rich, lingering head. Fresh hop aroma, against a sweetly malty and soft fruity and sulfury background. Bright, pleasant fresh-hop flavor carries through a taste that starts sweet and fruity, softens in the middle and ends with moderate lingering bitterness. Of only medium body, it is still a rich and well-balanced beer.

70 Shilling Special Pale Ale. OG 1.034, FG 1.010, 3.3% ABV. A dark pale ale, hoppier than usual for Scottish ales. Cask and keg conditioned, lightly primed.

60 Shilling Pale Ale (Light). OG 1.030, FG 1.010 to 11, 3% ABV. A very flavorful and well-balanced ale, with roast barley predominant. One of the last 60 shilling light ales to be found it Scotland, it constitutes only 5 percent of Maclays sales and is difficult to find.

Maclays Strong Ale is also Fowlers Wee Heavy, brewed by Belhaven. A porter has also been brewed recently.

Orkney Brewery. Quoyloo, Sandwich, Orkney, KW16 3LT, Scotland, (085 684) 802. Roger White.

Products

Skullsplitter Ale. OG 1.080, 8.5% ABV.
Raven Ale. OG 1.038, 3.8% ABV.

Rose Street Brewery. 55-57 Rose Street, Edinburgh, EH2, Scotland, (031) 220 1227. Brewer Ronald Borzucki's domain is Scotland's first and most successful brewpub, opened by the Alloa Brewery in 1983. Fifty percent of sales are on-premises, and the other 50 percent goes to six other Alloa pubs around Edinburgh. Borzucki brews twice each week. Sixty percent of his production is 80 shilling ale.

Ron uses Edme Export Bitter (unhopped) extract, along with roast malt extract in his 90 shilling ale. Fermentation with Alloa yeast is at 58 degrees F (18 degrees C), to control esters. The ales go into 18 gallon (Imperial) casks unfiltered with premixed finings. The casks are fitted with a plastic shive with a porous spile. The ales are served without blanket pressure at a 52 to 54 degrees F (11 to 12 degrees C) cellar temperature.

Products

90 Shilling. OG 1.055 to 56, FG 1.013 to 18. 75% Export Bitter Malt Extract, 25% Diamalt Extract. Very pungent Saxon and spicy Hallertauer hops. Stoutlike, dark, fruity and estery. Roast/burnt finish is medium-long, lightly-dry.

Auld Reekie 80 Shilling. OG 1.040 to 43, FG 1.010. Deep copper color. Diacetyl dominates the nose, mixed with green apples and sulfur. Sweetness predominates in the flavor, but butterscotch and the tang of hops round it out. Light bitterness lingers in the generally sweet finish. Light body.

Scottish & Newcastle. Head Office: S & N Breweries PLC, Abbey Brewery, 11 Holyrood Road, Edinburgh, Lothian, EH8 8YS, Scotland, (031) 556 2591. Fountain Brewery, Fountainbridge, Edinburgh, Lothian EH3 9YY, Scotland, (031) 229 9377.

An amalgam of breweries from successive rationalizations that included William Younger's, McEwan's & Newcastle, Berwick's Holyrood, Robert Younger's, Bernards, Morison's, Robert Deuchar's and most of the other Scots breweries that are only memories, with several English breweries as well, including Theakstons. Currently a 20,000 employee giant, Scottish & Newcastle is one of Britain's "Big Six."

By the 1890's William Younger and William McEwan were the two largest brewers in Scotland, had gone public, and were engaged in swallowing up smaller competitors. McEwan had spent five years with John Jeffrey at the Heriot Brewery in Grassmarket before founding the Fountain Brewery in 1856. He entered into the colonial trade with great success. His sister's son William (son of James Younger of Alloa, and yes, all these Youngers brewing beer gets confusing) succeeded him to the directorship of McEwan's in 1886. Further ties with Younger were cemented when McEwan left shares to both William and Robert Younger upon his death. McEwan's and Younger's merged in 1931 under Harry George Younger, of Belhaven House, Dunbar. He said that the merger was "not intended to alter the individual characteristics of the ales marketed by the respective companies, but [to make] available to each the technical and other resources of both."[11]

The present Fountain brewery is fitted with continental-type stainless steel closed fermenters. Gypsum is added to the brewing liquor for pH adjustment in the mash. Two to 3 percent roast barley in the grist gives color and flavor to the Scottish ales. A protein rest is used, with adjuncts added at saccharification. Northern Brewer hops are used in the kettle for 25 to 50 percent of the bitterness, with isomerized extracts providing the balance.

Scottish & Newcastle's yeast is a highly attenuative, traditional ale yeast, selected in the 1970s from the mixed strain that had been in use for a hundred years. Fermentation temperatures are limited to highs of 68 to 72 degrees F (20 to 22 degrees C), and oxygenation is limited in order to control yeast growth and ester formation. Ale batches are brewed at high gravity (OG 1.065 to 70), then diluted. Lagers are brewed at OG 1.070. S&N still owns one maltings, a modern plant at Moray Forth.

Brand names are applied variously to the products, depending on the intended market.

Products

Gordon's Highland Scotch Ale. OG 1.090, 8.5% ABV. Brewed in Edinburgh, then shipped in bulk for bottling in Belgium. Not sold in Scotland.

McEwan's Scotch Ale. OG 1.088, 8% ABV. Exported to the United States. Exported elsewhere as Younger's Double Century, and brewed under license in Belgium. Not sold in the United Kingdom.

Younger's No. 3. OG 1.043, 4.5% ABV.

McEwans 80 Shilling. OG 1.042, 4.5% ABV, 25 °EBC. The

brand leader in Scotland. Also bottled and kegged as McEwan's Export, and Younger's IPA. Translucent reddish-brown color. Dense head and rich lacing. Fruity and slightly hoppy and roasty aroma. Mild flavor, softly sweet, very slightly fruity. The soft and sweet finish is only faintly bitter. The flavors in the Export version are less full, being masked by carbonation.

McEwan's 70 Shilling. Also sold as Younger's Scotch Ale and Tartan Special. OG 1.036.5, 3.7% ABV, 22 °EBC. Light copper color. Well balanced ale with complex malt character, butteriness and subdued fruitiness. Slightly hoppy finish. Medium-light body and gassy.

Younger's Pale Ale. Very light copper color. Sharp hoppiness. Medium-light body, mildly sweet. Medium finish, very mildly bitter.

Younger's No. 3. OG 1.043. Rich and dark Scottish ale, brewed primarily for the English market.

Tennent Caledonian Breweries. Wellpark Brewery, 110 Bath Street, Glasgow, Strathclyde G2 2ET, Scotland, (041) 552 6552. Also Herriot Brewery, Roseburn Terrace, Edinburgh, Lothian, EH12 5LY, Scotland, (031) 337 1361. Essentially Bass' Scottish arm, Tennent brews mostly lager.

The brewery's history goes back to 1556. Records of that year show that Robert Tennent was a private brewer and maltster near Glasgow Cathedral, and an early member of the Incorporation of Maltmen. Another Robert Tennent brewed at his White Hart Inn in Glasgow in the mid-18th century. J & R Tennent began brewing in 1769 at their Drygate brewery, eventually expanding to the adjacent Wellpark brewery in 1793. Tennent's was exporting to expatriate tobacco planters in Virginia by 1797. Tennent's Wellpark Brewery was the largest in the west of Scotland in 1858. At that time their Scotch ales were world famous, and Tennent's had become the biggest beer exporters in the world.

Tennent's, like the other Scottish brewers, had reacted to the infringement of pale ale on their markets by adding that brew style to their portfolio in the mid-19th century. When German and Danish lagers began to capture a share of their markets in the latter half of the century, Tennent's became the first Scottish brewery to tackle the fizzier lagers. Tennent's began brewing lager in May, 1885. In 1891 they completed construction of a German-made brewery dedicated to lager production; Glasgow became the bastion of lager drinking in Great Britain.

The brewery continued to grow until the firm took over rival Maclachlan's in the early 1960s. The debt incurred by the takeover then forced them to sell out to Charrington United Breweries Limited, already the masters of United Caledonian Breweries (Aitken, Fowler, Calder, Murray, Jeffrey and George Younger). In January 1966, they were merged into Tennent Caledonian Breweries Ltd. The 1967 merger with Bass brought them under the control of Bass Charrington.

Tennent's claims the dubious distinction of having introduced lager brewing to Scotland (by the early 1980s lagers outsold ales in Scotland), and for introducing tinned (canned) beer to Britain. Their particularly tacky and not-politically-correct "Girls of Tennent's" label series sums up the company's outlook. The current brewery is equipped with state-of-the-art Steinecker and Balfour equipment.

Products

Tennent's 80 Shilling. OG 1.042, 4.2% ABV. Formerly an ale of great measure, it is still pleasant, if unexceptional. Difficult to find in casks.

Traquair House Brewery. Innerleithen, Peeblesshire, Borders EH44 6PW, Scotland, (0896) 830323. The brewery is in the oldest inhabited house in Scotland, restored in 1965 by 20th laird, Peter Maxwell Stewart (d.1990). His wife Flora Stewart survives him, and his daughter Lady Catherine Maxwell Stewart, the 21st Lady of Traquair, now manages Traquair House's affairs.

The manor has sheltered 27 Scottish (and English) kings and queens. The original tower dates from the 11th century. The rest of the structure all was built before 1700. Stewarts have been the Lairds of Traquair since 1478.

Laird Peter Maxwell resurrected the cellar brewery, which dates back to 1573, the time of the 4th Laird, Sir John Stuart, Captain of Mary Queen of Scots' bodyguard. Much of the brewery was intact in 1965, and the laird put it to work brewing using an 18th century recipe for strong ale. Ian Cameron started assisting him in the brewery 17 years ago, and has run the brewery himself for three years. Frank Smith now assists the capable Ian with mashing in, cleaning and deliveries.

The manor itself, strategically located in the Borders of southern Scotland, has a long and tempestuous history. "The Sleekit Yetts", or Bear Gates, were closed in the autumn of 1745 by the 5th Earl of Traquair upon the departure of Prince Charles Edward Stuart (Bonnie Prince Charlie) for England, with the vow that they would not be reopened until a Stuart again sat on the Scottish throne. Bonnie Prince Charlie and his army of volunteer Scots threatened London itself before they were defeated. The cause of Scottish independence lost its last war; the Bear Gates remain closed.

The four barrel (U.K.) brewery uses water of 100 milligrams per liter total dissolved solids, 75 milligrams per liter hardness and 11 milligrams per liter total alkalinity. The water is piped two and one half miles from the spring down to the brewery. The grist is composed of pale malt and 1 percent roast barley. After infusion mashing, the wort is run to the "new" copper, built in 1736 during the tenancy of the 11th laird, Charles Stewart. The wort is boiled 1 1/2 hours for Bear Ale, and four to five hours for the strong ale. The wort density rises SG one point every 15 minutes. Kettle caramelization has a major effect on Traquair House Ale's flavor. East Kent Goldings are added at the onset of boiling, and again thirteen minutes before final gravity is reached. For the Scotch Ale, Goldings are added in five to six pound and two to three pound increments, depending on their alpha acidity. The wort is run through a sieve, exactly as Scots brewers did two centuries ago, and into Coolship No. 1. Cooling in the wooden coolships, and thence over the Baudelot cooler obtained in 1984, takes 1 1/2 half hours.

Fermentation takes place in one of three unlined fermenters of Russian Memel oak (*Quercus robur* and *Quercus petraea* are the most common oak species that grow in the Baltic region. The wood is probably of one or both of these species). Two of the fermenting tuns date to 1811, the other "is not yet 30."[15] Attenuation starts as high as 68 degrees F (20 degrees C), but never rises above 70 degrees F (21 degrees C).

The yeast strain pitched is from another Scottish brewery. It ferments strongly, being skimmed twice on the second day, and fermenting out in 3 1/2 days. Temperature is controlled by opening or shutting a window, or running the beer through a copper line to cool it. The ales are left in the fermenters for a total of 10 days before being racked to 36 gallon (U.K.) barrels with isinglass finings. After 1 1/2 to two weeks, the ales go out to the trade. After three to four months aging in casks to develop its soft, round flavor, the Wee Heavy is sent to Belhaven for bottling. It is bottled 40 barrels (U.K.) at a time.

Total annual production of the Scotch Ale has never exceeded 210 barrels. In 1991, a meager 240 cases of

Traquair House Ale were destined for the United States. Overall, two-thirds of production is exported. The ale is not difficult to find in Edinburgh and the borders, but is less readily found elsewhere.

In 1984 Alan Eames imported a thousand cases to the United States; some of the bottles are still laid down in cellars here. Recently, Merchant du Vin has undertaken to market Traquair House Ale here in the United States for the Stewarts; this rarest of ales might even grace the shelves of your neighborhood store.

It is not unlikely that Bonnie Prince Charlie drank the ale in 1745, on his march toward London.

Products

Traquair House Ale. OG 1.075, FG 1.012 to 15, 8% ABV. Deep burnished-copper color. Redder than other Scotch ales. Amber and ruby tones shine brightly. Very full and round flavor. Great depth of maltiness is dominated by alcohol, hoppiness and hop bitterness. Caramel/burnt undertones. Short, dry finish. Harsh flavors in the young ale are not evident in more mature casks. Candylike flavor increases over time.

Bear Ale. OG 1.050, FG 1.010, 5.2% ABV. Strong draft ale. Medium copper color. Creamy head leaves a fine, dense, glass-coating lace. Primarily hoppy, with the bright, fresh taste of hops predominant. Sweet, clean malt flavor, tangy, with medium-light bitterness. Relatively soft, long finish without harshness.

Fair Ale. OG 1.042. Six barrels are brewed each year for the castle's huge summer festival. Very sweet and light, "a right summer drink."

Visitors to Traquair House are well advised to enjoy the hospitality at the Traquair Arms Hotel, Traquair Road off the A72, phone 83022. Bear Ale and Trauquair House Ale are among the four or five real ales on tap, or take a glass of mead from Moniak Castle, Inverness.

West Highland Breweries. Old Station Brewery, Taynuilt, Argyllshire, Strathclyde, PA35 1JB, Scotland, (086 62) 246. Richard Saunders.

Products

Highland Severe. OG 1.050.
Highland Heavy. OG 1.038.
Highland Dark Light (HDL). OG 1.034.

Campbell's Scotch Ale, bottled in Belgium is brewed by Courage. Campbell's was once a respected brand of the Campbell, Hope and King Brewery.

Appendix B: Edinburgh Pubs

There are many outstanding pubs throughout Scotland, but more Scottish Ales are available in and around Edinburgh than anywhere else in Scotland, although Maclays and Belhaven can be hard to find. Most of the pubs are tied houses, which means that they offer little choice of ales other than those brewed or marketed by the brewery to which they are tied. If there is an Alloa Ales sign outside, Inde Coope Burton Ale is likely to be served as well, but don't expect to find Maclays on tap! Some of the free houses offer a wide selection of real ales. Good pubs are to be found clustered on Rose Street and in Grassmarket/Cowgate, as well as on George Street and along the Royal Mile. Pub ownership, offerings and atmosphere are likely to change with time, and the traveler in Scotland would be well advised to obtain a copy of CAMRA's Scottish Real Beer Guide for current information, including which ales are on tap.

The Athletic Arms—Better known as Diggers or the Gravediggers. The most visible sign says A. Thomas Wilkie, Wine, Spirits & Ale Merchants. Their cellarman's skill is said to produce the best pint of 80 shilling in Scotland. The pub

Kenilworth pub in Edinburgh. Photo by Greg Noonan.

is home to the fans of the Hearts of Midlothian Football Club, whose playing field is nearby. Scottish & Newcastle. At the corner of Angle Park Terrace and Henderson Terrace.

Bailie Bar—One-half flight down. Belhaven. 2 St. Stephen Street.

Bannerman's Bar—Housed in a medieval vault, this pub is popular with students. Alloa. 212 Cowgate.

The Bull—An unassuming pub with a good selection of real ales. 12 Grassmarket.

Forrest Hill Bar—Also known as Sandy Bell's. One of the better-known of the 50-odd pubs in the Old Town. Scottish & Newcastle. 23 Forrest Road.

Golf Tavern—Reputed to be the oldest pub in Edinburgh. Elegant free house. Wright's House, Bruntsfield Road.

Guilford Arms—Call. If they have Traquair House Ale on draft, go. East end of Princes Street, near West Register Street.

The Kenilworth—A comfortable, traditional pub with a younger clientele. Alloa. Rose Street.

Malt and Shovel—A small chain of pubs featuring single malts and good selections of real ales, sometimes even Traquair House Ale.

Milnes—Basement pub that has been haunted by generations of Scottish artists. Scottish & Newcastle. Corner of Hanover and Rose Streets.

Starbuck Tavern—Also known as Scott's. Cozy pub with a view of the Firth of Forth estuary. Belhaven 90 shilling, 80 shilling, 70 shilling and 60 shilling on draft. Grantin Way, Laverock Bank Road, Newhaven.

T.G. Willis—Upstairs bar. Noted for traditional Scottish fare and a wide range of cask ales, including Caledonian, Belhaven and Broughton. George Street.

The Tilted Wig—Upscale pub in the New Town. Maclays. 3 Cumberland Street.

The Victoria and Albert—A sing-along dinner house, housed in a beautiful pub. Alloa. Frederick Street.

The Volunteer Arms— Also known as The Canny Man. Another famous pub, family owned since 1871. Caledonian and others. Morningside Road.

Appendix C:
Weights and Measures

DRY MEASURE

Bushel, Imperial	= 2218.192 cubic inches = 42 pounds of malt = 56 pounds of barley
Bushel, Winchester	= 2150.42 cubic inches = same measure as U.S. bushel
Quarter	= 8 bushels = 336 pounds of malt = 448 pounds of barley

Scotland before 1840:

Bushel	= volume of 10 Imperial gallons of water, but in practice variable.
	= 4 pecks = 8 stimperts = 16 lippies
Calder	= 2,240 pounds of grain = 16 bolls
Boll	= 140 pounds of grain = 4 firlots = 2 auchlets

LIQUID MEASURE

Mutchkin, Old Scottish	= 1/4 pint
Chopin, Old Scottish	= 1/2 pint
Pint, Old Scottish	= 3 Imperial pints = 3.6 U.S. pints

	= 1.71 liters
Pint, U.S.	= 16 fluid ounces = 0.8327 Imperial pints
	= 0.4732 liter
Quart, Old Scottish	= 2 Old Scottish pints
Quart, Imperial	= 1.2 U.S. quarts = 1.1359 liters
Quart, U.S.	= 0.833 Imperial quarts = 0.9464 liter
Gallon, Imperial	= 1.2009 U.S. gallons = 4.5436 liters
Gallon, Old English Beer/Ale	= 282 cubic inches = 4.62 liters
Gallon, Old English Wine	= 231 cubic inches, same as U.S. gallon
Pin	= 4 1/2 U.S. gallons
Firkin	= 9 U.S. gallons
Kilderkin	= 18 U.S. gallons
Barrel, U.K.	= 36 Imperial gallons = 43.23 U.S. gal.
	= 163.65 liters
Hogshead, Old English	= 51 gallons of ale or beer = 1 1/2 U.K. bbl.
Hogshead, London	= 54 gallons of beer = 48 gallons of wine
Hogshead, Imperial measure	= 52.5 Imperial gallons = 63 U.S. gallons
Puncheon, Old English	= 80 Imperial gallons = 96.1 U.S. gallons
Puncheon	= 70 Imperial gallons = 84 U.S. gallons
Butt	= 108 Imperial gallons
Tun	= 252 wine gallons

WEIGHTS

Ounce (avoirdupois)	= 28.3495 grams
Pound (avoirdupois)	= 453.5924 grams = 0.4536 kilograms

TEMPERATURE CONVERSIONS

Degrees Fahrenheit to degrees Centigrade	°C = °F – 32 ÷ 1.8
Degrees Centigrade to degrees Fahrenheit	°F = °C x 1.8 + 32

MONEY

Shilling, Old English	= 12 pence = 133.2 grams of silver, notated s or /
Shilling, after 1930	= 12 pence = 80.73 grams of silver, notated s or /
Pound, Old English	= 240 pence = 20 shillings , notated £
Pound, since 1971	= 100 pence, notated £
Guinea, 1663-1813	= 21 shillings
Penny	= notated d (for the Roman coin, a denarius)
Penny, since 1971	= notated p

Scottish money was originally identical in value with English, but depreciated during the wars with England. By the 16th century the Scottish shilling was only valued at one English penny. In 1707 Scottish money was standardized to English values again.

USEFUL INFORMATION

Pounds Per Barrel	= weight of 36 gallons (Imperial) of wort minus weight of 36 gallons of distilled water = 0.3336 grams per 100 milliliters
1 pound of sucrose	= OG 1.036 in an Imperial gallon = OG 1.043 in a U.S. gallon displaces 1/16 of an Imperial gallon of water
1 pound of malt extract	= OG 1.035-1.040 in a U.S. gallon
1 pounds of dry malt	= OG 1.040-1.045 in a U.S. gallon

In the 1820s, 50 shilling	Ale was OG 1.080-86,	FG 1.032-35	5.7% ABW	
70 shilling	1.090-95	1.036-39	6.4%	
80 shilling	1.100-08	1.040-44	7.0%	
90 shilling	1.110-16	1.045-47	7.5%	
110 shilling	1.120-25	1.048-50	8.3%	
Late 1830s, 40 shilling	1.078	1.025	5.5%	
50 shilling	1.090	1.030	6.5%	
70 shilling	1.105	1.042	7.5%	
80 shilling	1.120	1.045	8.5%	
5 guinea	1.125	1.060	8.5%	

Glossary

acetification. The changes brought about by production of acetic acid, generally as spoilage by aerobic bacteria, but also as mash or kettle pH adjustment.

adjunct. Any *unmalted* grain or other fermentable ingredient added to the mash.

aeration. The action of introducing air to the wort at various stages of the brewing process.

airlock. (see fermentation lock)

airspace. (see ullage)

alcohol by volume (v/v). The percentage of volume of alcohol per volume of beer. To calculate the approximate volumetric alcohol content, subtract the terminal gravity from the original gravity and divide the result by 75. For example: 1.050 - 1.012 = .038 / 75 = 5% v/v.

alcohol by weight (w/v). The percentage weight of alcohol per volume of beer. For example: 3.2% alcohol by weight = 3.2 grams of alcohol per 100 centiliters of beer.

ale. 1. Historically, an unhopped malt beverage. 2. Now a generic term for hopped beers produced by top fermentation, as opposed to lagers, which are produced by bottom fermentation.

all-extract beer. A beer made with only malt extract as opposed to one made from barley, or a combination of malt extract and barley.

all-grain beer. A beer made with only malted barley as opposed to one made from malt extract, or from malt extract and malted barley.

all-malt beer. A beer made with only barley malt with no adjuncts or refined sugars.

alpha acid. A soft resin in hop cones. When boiled, alpha acids are converted to iso-alpha-acids, which account for 60 percent of a beer's bitterness.

alpha-acid unit. A measurement of the potential bitterness of hops, expressed by their percentage of alpha acid. Low = 2 to 4%, medium = 5 to 7%, high = 8 to 12%. Abbrev: A.A.U.

attenuation. The reduction in the wort's specific gravity caused by the transformation of sugars into alcohol and carbon-dioxide gas.

Balling. A saccharometer invented by Carl Joseph Napoleon Balling in 1843. It is calibrated for 63.5 degrees F (17.5 degrees C), and graduated in grams per hundred, giving a direct reading of the percentage of extract by weight per 100 grams solution. For example: 10 °B = 10 grams of sugar per 100 grams of wort.

Bitterness Units (BU). ASBC measurement of bittering substances in beer, primarily iso-alpha-acids, but also including oxidized beta acids. Also see International bitterness units.

blow-by (blow-off). A single-stage homebrewing fermentation method in which a plastic tube is fitted into the mouth of a carboy, with the other end submerged in a pail of sterile water. Unwanted residues and carbon dioxide are expelled through the tube, while air is prevented from coming into contact with the fermenting beer, thus avoiding contamination.

carbonation. The process of introducing carbon-dioxide gas into a liquid by: 1. injecting the finished beer with carbon dioxide; 2. adding young fermenting beer to finished beer for a renewed fermentation (kraeusening); 3. priming (adding sugar) to fermented wort prior to bottling, creating a secondary fermentation in the bottle.

carboy. A large glass, plastic or earthenware bottle.

cast. As in "the wort is cast from the kettle"; the running-off of the wort after processing.

chill haze. Haziness caused by protein and tannin during the secondary fermentation.

culm. Dusty or inferior anthracite coal; a poor fuel for kilning.

dry hopping. The addition of hops to the primary fermenter, the secondary fermenter, or to casked beer to add aroma and hop character to the finished beer without adding significant bitterness.

dry malt. Malt extract in powdered form.

dusty yeast. Yeast which does not quickly precipitate out of suspension at the end of fermentation.

EBC (**European Brewery Convention**). (see SRM)

extract. The amount of dissolved materials in the wort after mashing and lautering malted barley and/or malt adjuncts such as corn and rice.

falling heat. The temperature of sweet wort run from the mash tun.

fecula. Trub.

fermentation lock. A one-way valve, which allows carbon-dioxide gas to escape from the fermenter while excluding contaminants.

final specific gravity. The specific gravity of a beer when fermentation is complete.

fining. The process of adding clarifying agents to beer during secondary fermentation to precipitate suspended matter.

flatten. Conditioning under atmospheric pressure, to release carbon dioxide from the beer.

flocculant yeast. Yeast which forms large colonies and tends to come out of suspension before the end of fermentation.

flocculation. The behavior of yeast cells joining into masses and settling out toward the end of fermentation.

flooring the pieces. Spreading the barley onto the malting floor to germinate (malt).

free house. A pub (public house) which is not owned by a brewery company. A free house is free to purchase any beer from any brewery that will sell it, consequently, many free houses have a wider selection.

gyle. Fermenting beer, or the tun in which fermentation takes place.

homebrew bittering units. A formula invented by the American Homebrewers Association to measure bitterness of beer. Example: 1.5 ounces of hops at 10 percent alpha acid for five gallons: 1.5 x 10 = 15 HBU per five gallons.

hop pellets. Finely powdered hop cones compressed into tablets. Hop pellets are 20 to 30 percent more bitter by weight than the same variety in loose form.

hydrometer. A glass instrument used to measure the specific gravity of liquids as compared to water, consisting of a graduated stem resting on a weighed float.

Independent Tied House Chains. A company which is not a brewery may own a chain of pubs. The individual pubs are tied to the chain, which controls what brand of beer is served. But since the pubs are not owned by a brewery company, they are technically under the licensing laws "independent."

International bitterness units. The EBC measurement of the concentration of iso-alpha-acids in milligrams per liter (parts per million) in wort and beer. Also see Bitterness Units.

isinglass. A gelatinous substance made from the swim bladder of certain fish and added to beer as a fining agent.

kraeusen. (n.) The rocky head of foam which appears on the surface of the wort during fermentation. (v.) To add fermenting wort to fermented beer to induce carbonation through a secondary fermentation.

lager. (n.) A generic term for any bottom-fermented beer. Lager brewing is now the predominant brewing method worldwide except in Britain where top fermented ales dominate. (v.) To store beer at near-zero temperatures in order to precipitate yeast cells and proteins and improve taste.

lauter tun. A vessel in which the mash settles and the grains are removed from the sweet wort through a straining process. It has a false, slotted bottom and spigot.

Limewood. Wood of the linden tree (*Tilia sp.*) known in England as lime tree and in Scotland as lorit.

liquefaction. The process by which alpha-amylase enzymes degrade soluble starch into dextrin.

malt. Barley that has been steeped in water, germinated, then dried in kilns. This process converts insoluble starchs to soluble substances and sugars.

malt extract. A thick syrup or dry powder prepared from malt.

mashing. Mixing ground malt with water to extract the fermentables, degrade haze-forming proteins and convert grain starches to fermentable sugars and nonfermentable carbohydrates.

maskin' loom. Mash tun.

modification. 1. The physical and chemical changes in barley as a result of malting. 2. The degree to which these changes have occured, as determined by the growth of the acrospire.

original gravity. The specific gravity of wort previous to fermentation. A measure of the total amount of dissolved solids in wort.

pH. A measure of acidity or alkalinity of a solution, usually on a scale of one to 14, where seven is neutral.

pickle. The kernel of barley or malt.

pieces. Barley from the steeps, laid on a malting floor to germinate.

Plato. A saccharometer that expresses specific gravity as extract weight in a one-hundred-gram solution at 68 degrees F (20 degrees C). A revised, more accurate version of Balling, developed by Dr. Plato.

primary fermentation. The first stage of fermentation, during which most fermentable sugars are converted to ethyl alcohol and carbon dioxide.

priming sugar. A small amount of corn, malt or cane sugar added to bulk beer prior to racking or at bottling to induce a new fermentation and create carbonation.

puncheon. A cask, usually employed for conditioning, of 70 Imperial gallons, (84 U.S. gallons).

racking. The process of transferring beer from one container to another, especially into the final package (bottles, kegs, etc.).

saccharification. The naturally occurring process in which malt starch is converted into fermentable sugars, primarily maltose.

saccharometer. An instrument that determines the sugar concentration of a solution by measuring the specific gravity.

secondary fermentation. 1. The second, slower stage of fermentation, lasting from a few weeks to many months depending on the type of beer. 2. A fermentation occuring in bottles or casks and initiated by priming or by adding yeast.

shive. The bung, or cork, of a beer barrel, that is pierced for a spile.

sowens. 17th century Scots name for the hopped beers of Denmark.

sparging. Spraying the spent grains in the mash with hot water to retrieve the remaining malt sugar.

specific gravity. A measure of a substance's density as compared to that of water, which is given the value of 1.000 at 39.2 degrees F (4 degrees C). Specific gravity has no accompanying units, because it is expressed as a ratio.

spile. A wooden spear about two inches long. It is fitted into the shive sealing a cask of ale/beer. A soft spile allows pressure to vent; a hard spile reseals the cask.

SRM (Standard Reference Method) and EBC (European Brewery Convention). Two different analytical methods of describing color developed by comparing color samples. Degrees SRM, approximately equivalent to degrees Lovibond, are used by the ASBC (American Society of Brewing Chemists) while degrees EBC are European units. The following equations show approximate conversions:

(°EBC) = 2.65 x (° Lovibond) – 1.2

(° Lovibond) = 0.377 x (°EBC) + 0.45

starter. A batch of fermenting yeast, added to the wort to initiate fermentation.

stillion. The rack upon which casks, hogsheads or puncheons are placed. Also, the hogsheads themselves when their contents are conditioning.

strike temperature. The initial temperature of the water when the malted barley is added to it to create the mash.

tied house. A pub (public house) owned by or in debt to, a brewery, which consequently controls what brand of beer is served. Until 1991, most British pubs were tied houses.

torrified wheat. Wheat which has been heated quickly at high temperature, causing it to puff up, which renders it more easily mashed.

trub. Suspended particles resulting from the precipitation of proteins, hop oils and tannins during boiling and cooling stages of brewing.

tun. Any open tank or vessel. More usually applied to a mashing tub, but until the 19th century commonly used in reference to a fermenting vessel.

ullage. The empty space between a liquid and the top of its container. Also called airspace or headspace.

v/v. (see alcohol by volume)

w/v. (see alcohol by weight)

water hardness. The degree of dissolved minerals in water.

wort. The mixture that results from mashing the malt and boiling the hops, before it is fermented into beer.

Index

Bibliography

LISTED BY NUMBER

1. Charles Mc Master. 1985. "Alloa Ale." Alloa advertising, 1827. *Scottish Brewing Archives*, Edinburgh.

2. Rev. George Gilfillan. (ed.) 1856. *Poetical Works of Robert Burns*. Appleton & Co., New York.

3. John Harrison. et al. 1991. *Old British Beers and How to Make Them*. Durden Park Beer Circle, London.

4. W.H. Roberts. 1847. *The Scottish Ale Brewer and Practical Maltster*. A. & C. Black, Edinburgh.

5. David Johnstone. March, 1984. *In Search of Scotch Ale*. Institute of Brewing, Irish Section, London.

6. David Johnstone. July, 1983. "One Hundred Years of Lager Brewing in Scotland." *The Brewer*.

7. David Johnstone. 1989. *The World Beer Market—Trends, Types and Technology.* Lecture given for International Brewing Science and Technology Training course at Heriot-Watt University, Edinburgh.

8. Charles Mc Master. 1985. "Alloa Ale." *Scottish Brewing Archives*. Edinburgh.

9. J. Bickerdyke. 1886. *The Curiosities of Ale and Beer.* Field & Tuer, London.

10. Alfred Barnard. 1889-90. *The Noted Breweries of Great Britain and Ireland.* Sir Joseph Causton & Sons, London.

11. Ian Donnachie. 1979. A History of Brewing in Scotland. John Donald Publishers Ltd., Edinburgh.

12. Charles Mc Master. 1988. "Porter Brewing in Scotland." *Scottish Brewing Archive,* No. 12. Edinburgh.

13. Charles Mc Master. 1989. "A Short History of Slateford Maltings." *Scottish Brewing Archive,* No. 14. Edinburgh.

14. Charles Mc Master. 1988. "Scotland's Forgotten Breweries: The Boroughloch Brewery." *Scottish Brewing Archive* No. 11. Edinburgh.

15. Personal communication to the author.

16. *Scottish Brewing Archives*. Holdings. Edinburgh.

17. Clive La Pensee. 1990. *The Historical Companion to House-Brewing.* Montag Publications, Beverly.

18. Charles Mc Master. 1990. *Scottish Brewing Archive*, No. 17. Edinburgh.

19. Ian Donnachie. 1985. "Men of Brewing: Viscount Younger of Leckie." *Scottish Brewing Archive*, No. 6. Edinburgh.

20. John Jeffrey. 1875. Well book, Vol. 13, No. 3. in *Scottish Brewing Archives*. Holdings. Edinburgh.

21. Stephen Cribb. 1990. "Beer and Rocks." ***zymurgy*** Fall 1990 (Vol. 13, No. 3). American Homebrewers Association, Boulder, Colo.

22. Charles Mc Master. 1989-90. "Nemesis on the Pleasance." *Scottish Brewing Archive*, No. 16. Edinburgh.

23. A. Morrice. 1827. *A Practical Treatise on Brewing the Various Sorts of Malt Liquor* quoted in Donnachie, Jan. 1979, *A History of Brewing in Scotland*. John Donald Publishers, Ltd., Edinburgh.

24. K.C. Lam., R.T. Foster II and M.L. Deinzer 1986. "Aging of Hops and Their Contribution to Beer Flavor." *Journal of Agricultural Food Chemistry*. Vol. 34, No. 4.

25. J.S. Hough, D.E. Briggs and R. Stevens. 1982. *Malting and Brewing Science*. (Vol. 2). Chapman and Hall, London.

26. CAMRA. 1989. *A Guide to Good Beer in the Forth Valley, Stirling and the Trossachs*. CAMRA, St. Albans.

27. K. Dean. 1988. "Ballingall of Dundee." *Scottish Brewing Archive*, No. 11. Edinburgh.

28. Ian Donnachie. 1985. "Men of Brewing: William McEwan." *Scottish Brewing Archive,* No. 5. Edinburgh.

29. T.D. Kane and Charles Mc Master. 1989. "Croft-An-Righ Brewery." Scottis*h Brewing Archive,* No. 14. Edinburgh.

30. Charles Mc Master. 1990-91. *Scottish Brewing Archive,* No. 18. Edinburgh.

31. T. Rutherford. 1985. "Prohibition—The Brewers' Response." *Scottish Brewing Archive,* No. 6. Edinburgh.

32. Peter Maxwell Stewart. *Traquair House, A Historical Survey.* Innerleithen.

LISTED BY AUTHOR

Alfred Barnard. 1889-90. *The Noted Breweries of Great Britain and Ireland.* Sir Joseph Causton & Sons, London. (10)

J. Bickerdyke. 1886. The Curiosities of Ale and Beer. Field & Tuer, London. (9)

CAMRA. 1989. *A Guide to Good Beer in the Forth Valley, Stirling and the Trossachs.* CAMRA, St. Albans. (26)

Stephen Cribb. 1990. "Beer and Rocks." ***zymurgy*** Fall 1990 (Vol. 13, No. 3). American Homebrewers Association, Boulder, Colo. (21)

K. Dean. 1988. "Ballingall of Dundee." *Scottish Brewing Archive,* No. 11. Edinburgh. (27)

Ian Donnachie. 1979. A History of Brewing in Scotland. John Donald Publishers Ltd, Edinburgh. (11)

Ian Donnachie. 1985. "Men of Brewing: Viscount Younger of Leckie." *Scottish Brewing Archive,* No. 6. Edinburgh. (19)

Ian Donnachie. 1985. "Men of Brewing: William McEwan." *Scottish Brewing Archive,* No. 5. Edinburgh. (28)

Rev. George Gilfillan. (ed.) 1856. Poetical Works of Robert Burns. Appleton & Co., New York. (2)

John Harrison. et al. 1991. *Old British Beers and How to Make Them.* Durden Park Beer Circle, London. (3)

J.S. Hough, D.E. Briggs and R. Stevens. 1982. *Malting and Brewing Science.* (Vol. 2). Chapman and Hall, London. (25)

John Jeffrey. 1875. Well book, Vol. 13, No. 3. in *Scottish Brewing Archives.* Holdings. Edinburgh. (20)

David Johnstone. July, 1983. "One Hundred Years of Lager Brewing in Scotland." *The Brewer.* (6)

David Johnstone. March, 1984. In Search of *Scotch Ale.* Institute of Brewing, Irish Section, London. (5)

David Johnstone. 1989. *The World Beer Market—Trends, Types and Technology.* Lecture given for International Brewing Science and Technology Training course at Heriot-Watt University, Edinburgh. (7)

T.D. Kane and Charles Mc Master. 1989. "Croft-An-Righ Brewery." *Scottish Brewing Archive,* No. 14. Edinburgh. (29)

K.C. Lam, R.T. Foster II and M.L. Deinzer. 1986. "Aging of Hops and Their Contribution to Beer Flavor." *Journal of Agricultural Food Chemistry*. Vol. 34, No. 4. (24)

Clive La Pensee. 1990. *The Historical Companion to House-Brewing*. Montag Publications, Beverly. (17)

Charles Mc Master. 1985. "Alloa Ale." *Scottish Brewing Archives*. Edinburgh (8)

Charles Mc Master. 1985. "Alloa Ale." Alloa advertising, 1827. *Scottish Brewing Archives*, Edinburgh. (1)

Charles Mc Master. 1989-90. "Nemesis on the Pleasance." *Scottish Brewing Archive*, No. 16. Edinburgh. (22)

Charles Mc Master. 1988. "Porter Brewing in Scotland." *Scottish Brewing Archive*, No. 12. Edinburgh. (12)

Charles Mc Master. 1988. "Scotland's Forgotten Breweries: The Boroughloch Brewery." *Scottish Brewing Archive*, No. 11. Edinburgh. (14)

Charles Mc Master. 1990. *Scottish Brewing Archive*, No. 17. Edinburgh. (18)

Charles Mc Master. 1990-91. *Scottish Brewing Archive*, No. 18. Edinburgh. (30)

Charles Mc Master. 1989. "A Short History of Slateford Maltings." *Scottish Brewing Archive*, No. 14. Edinburgh. (13)

A. Morrice. 1827. *A Practical Treatise on Brewing the Various Sorts of Malt Liquor* quoted in Donnachie, Jan. 1979, *A*

History of Brewing in Scotland. John Donald Publishers, Ltd., Edinburgh. (23)

Personal communication to the author. (15)

W.H. Roberts. 1847. The Scottish Ale Brewer and Practical *Maltster.* A. & C. Black, Edinburgh. (4)

T. Rutherford. 1985. "Prohibition—The Brewers' Response." *Scottish Brewing Archive,* No. 6. Edinburgh. (31)

Scottish Brewing Archives. Holdings. Edinburgh. (16)

Peter Maxwell Stewart. *Traquair House, A Historical Survey.* Innerleithen. (32)

BREWERS PUBLICATIONS ORDER FORM

GENERAL BEER AND BREWING INFORMATION

QTY.	TITLE	STOCK #	PRICE	EXT. PRICE
______	The Art of Cidermaking	468	9.95	________
______	Brewing Mead	461	11.95	________
______	Dictionary of Beer and Brewing	462	19.95	________
______	Evaluating Beer	465	19.95	________
______	Great American Beer Cookbook	466	24.95	________
______	New Brewing Lager Beer	469	14.95	________
______	Victory Beer Recipes	467	11.95	________
______	Winners Circle	464	11.95	________

CLASSIC BEER STYLE SERIES

QTY.	TITLE	STOCK #	PRICE	EXT. PRICE
______	Pale Ale	401	11.95	________
______	Continental Pilsener	402	11.95	________
______	Lambic	403	11.95	________
______	Oktoberfest, Vienna, Märzen	404	11.95	________
______	Porter	405	11.95	________
______	Belgian Ale	406	11.95	________
______	German Wheat Beer	407	11.95	________
______	Scotch Ale	408	11.95	________
______	Bock	409	11.95	________
______	Stout	410	11.95	________

PROFESSIONAL BREWING BOOKS

QTY.	TITLE	STOCK #	PRICE	EXT. PRICE
______	Brewery Planner	500	80.00	________
______	North American Brewers Resource Directory	506	100.00	________
______	Principles of Brewing Science	463	29.95	________

THE BREWERY OPERATIONS SERIES, Transcripts From National Micro- and Pubbrewers Conferences

QTY.	TITLE	STOCK #	PRICE	EXT. PRICE
______	Volume 6, 1989 Conference	536	25.95	________
______	Volume 7, 1990 Conference	537	25.95	________
______	Volume 8, 1991 Conference, Brewing Under Adversity	538	25.95	________
______	Volume 9, 1992 Conference, Quality Brewing — Share the Experience	539	25.95	________

BEER AND BREWING SERIES, Transcripts From National Homebrewers Conferences

QTY.	TITLE	STOCK #	PRICE	EXT. PRICE
______	Volume 8, 1988 Conference	448	21.95	________
______	Volume 10, 1990 Conference	450	21.95	________
______	Volume 11, 1991 Conference, Brew Free or Die!	451	21.95	________
______	Volume 12, 1992 Conference, Just Brew It!	452	21.95	________

SUBTOTAL ________

Colo. Residents Add 3% Sales Tax ________

P&H * ________

TOTAL ________

Call or write for a free Beer Enthusiast catalog today.

- U.S. funds only.
- All Brewers Publications books come with a money-back guarantee.

***Postage & Handling:** $4 for the first book ordered, plus $1 for each book thereafter. Canadian and international orders please add $5 for the first book and $2 for each book thereafter. Orders cannot be shipped without appropriate P&H.

Brewers Publications, PO Box 1510, Boulder, CO 80306-1510, U.S.A.; (303) 546-6514; FAX (303) 447-2825

SCA